Active Learning Guide for College Physics

Volume 2

Eugenia Etkina

RUTGERS UNIVERSITY

Michael Gentile

RUTGERS UNIVERSITY

Alan Van Heuvelen

RUTGERS UNIVERSITY

PEARSON

BOSTON   COLUMBUS   INDIANAPOLIS   NEW YORK   SAN FRANCISCO   UPPER SADDLE RIVER
AMSTERDAM   CAPE TOWN   DUBAI   LONDON   MADRID   MILAN   MUNICH   PARIS   MONTRÉAL   TORONTO
DELHI   MEXICO CITY   SÃO PAULO   SYDNEY   HONG KONG   SEOUL   SINGAPORE   TAIPEI   TOKYO

Publisher: Jim Smith

Project Managers: Katie Conley and Beth Collins

Director of Development: Laura Kenney

Managing Development Editor: Cathy Murphy

Editorial Assistant: Kyle Doctor

Team Lead, Program Management, Physical
    Sciences: Corinne Benson

Director of Production: Erin Gregg

Full-Service Production and Composition:
    PreMediaGlobal

Illustrators: PreMediaGlobal

Manufacturing Buyer: Jeff Sargent

Marketing Manager: Will Moore

Cover Designer: Riezebos-Holzbaur Design Group
    and Seventeenth Street Studios

Cover Photo Credit: © Markus Altmann/Corbis

2 3 4 5 6 7 8 9 10—EBM—17 16 15 14 13

www.pearsonhighered.com

ISBN 10: 0-321-87711-X
ISBN 13: 978-0-321-87711-6

CONTENTS

PREFACE TO THE STUDENT

Welcome to college physics! If you are reading this preface, it is probably because your physics instructor has assigned the *Active Learning Guide,* a workbook to accompany the textbook *College Physics,* by Eugenia Etkina, Michael Gentile, and Alan Van Heuvelen. This workbook consists of carefully-crafted activities that supplement the textbook and provide an opportunity for further observation, sketching, analysis, and testing.

The chapters in this workbook correlate to those in the text. You will find marginal "Active Learning Guide" icons throughout the text that indicate content for which a workbook activity is available. Whether assigned to you or not, you can always use this workbook to reinforce the concepts you have read about in the text, to practice applying the concepts to real-world scenarios, or to work with sketches, diagrams, and graphs that help you visualize the physics.

The exercises in the *Active Learning Guide* are designed to help you learn to connect your intuition and everyday experiences with the words, graphs, and equations used in physics. In addition, we recommend that you apply these tips for success:

1. Actively participate in your own learning. Research shows that knowledge cannot be transmitted from one individual to another. It has to be actively constructed by the learner. This construction happens when you connect new ideas to what you already know.

2. Learn to think and act like a scientist. Be critical. Evaluate every piece of information and every new idea. Understand what is a result of observation and what is a result of reasoning. If you developed a new idea through reasoning, think of how you can test it in new experiments.

3. Reflect on the processes you use to construct and apply physics concepts. The logic you develop in physics will help you in your other courses and throughout your career.

4. Connect what you learn to the world around you. Sometimes what you learn may contradict what you thought you knew. Do not be discouraged. Great scientists such as Newton and Galileo also had to look at the world from a new point of view. This new perspective can make your study of physics more interesting and give you a deeper understanding of the world around you.

5. Learn to represent concepts in different ways to deepen your understanding. Physics uses many different ways to represent knowledge—pictures, diagrams, words, graphs, and equations. Memorizing formulas is not the key to success.

6. Learn to explain your ideas in simple language. The language that you use shapes your thinking; thus, being clear in your language will help you be clear in your thinking.

7. Pay attention to *how* you learn. You may understand better if you draw a picture or represent a process with a diagram. Perhaps you need to go over things several times. Maybe you need to check your algebra to avoid mistakes. Because people learn in different ways, this workbook employs many ways to construct physics knowledge. Knowing what helps you learn and apply knowledge most effectively will save you time and improve your work in physics and other classes.

8. Work collaboratively with friends. Research shows that students working together in study groups can solve problems that none of them can solve on their own. Through this collaboration, they can increase their skills individually.

9. Be patient. Learning physics requires focus and practice, just like learning a new sport or a musical instrument. As you become more proficient in physics, you will enjoy the learning process more and become more adept at applying your knowledge. Eventually, you will see physics all around you: in the bubbles rising from a pot of boiling water, in the behavior of a shower curtain when you take a shower, and in the glow of a filament in a lightbulb. Learning physics is empowering: it allows you to explain the world!

PREFACE TO THE INSTRUCTOR

Learning physics is easier for students when they actively observe, explain, test, represent, and evaluate explanations of physical phenomena they encounter in their everyday experiences. That understanding is the basis for the textbook *College Physics,* by Eugenia Etkina, Michael Gentile, and Alan Van Heuvelen, as well as this workbook, which accompanies that textbook.

Both the textbook and the *Active Learning Guide* are designed to encourage students in their study of physics by actively engaging them in the discovery process. The textbook includes Observational Experiment Tables and Testing Experiment Tables, as well as a problem-solving approach that will help students develop the skills they need to work through the assigned homework.

This workbook includes a set of activities for use in a variety of settings—large enrollment lectures and recitations, small classes, labs, and homework. You can use the activities in conjunction with the end-of-chapter problems in the textbook and/or *MasteringPhysics*, or on their own. Each chapter includes many activities. You do not have to assign them all. The *Instructor's Guide* that accompanies *College Physics* describes options for assigning the various activities. Each chapter of the textbook includes icons in the margins identifying where *Active Learning Guide* activities correspond to chapter content.

Although the *Active Learning Guide* is based on an active, process-oriented learning system, the activities will fit easily into any introductory physics course, regardless of the teaching method.

Within the *Active Learning Guide* are four categories of activities, each subdivided into activity types that help students develop specific cognitive and science-process abilities for learning and applying physics concepts. Like the textbook, each workbook chapter begins with qualitative concept building before introducing quantitative reasoning. Readers familiar with the original edition of the *Active Learning Guide* will recognize some activities and find many new ones.

1. Qualitative Concept Building and Testing activities help students construct new qualitative concepts by observing phenomena, recording their observations, devising qualitative explanations for their observations, and testing these explanations by using them to predict the outcome of testing experiments. These activities involve quantitative reasoning but no formal mathematics.

2. Conceptual Reasoning activities help students learn to reason about the physical world using the qualitative explanations that they have already tested and accepted. The students learn to represent phenomena in a number of different ways—through motion diagrams, force diagrams, work-energy bar charts, sketches, ray diagrams, and so forth—and in this way create referents that enhance their understanding of the more abstract physics quantities and concepts. These qualitative representations improve students' ability to reason about the world without using mathematics. Even instructors who prefer to introduce concepts differently from the method used in the Qualitative Concept Building and Testing activities will find these Conceptual Reasoning activities useful in helping students further their conceptual reasoning skills.

3. Quantitative Concept Building and Testing activities help students devise relationships between physical quantities based on the data that they collect or that are provided with the activity. Students then use the relationship to predict the outcome of a new experiment, which we call a *testing experiment*. It is important that students both participate in and reflect on this process. Reflection will improve their retention of ideas and allow them to see patterns in how they construct knowledge.

4. Quantitative Reasoning activities help students learn to use different types of representations to describe physical processes and to solve problems. Students use verbal, pictorial, diagrammatic, graphical, and mathematical representations to solve problems. Special evaluation activities help students learn to check for the consistency of their representations and to evaluate the correctness of their work. Examples of more complex problems and experimental design activities that can be used in labs are also provided. All activities within this section can be used effectively with various instructional methods.

The ***Instructor's Guide*** that accompanies ***College Physics*** will help you make the transition to this approach from the materials you have in prior classes. That guide includes suggestions for how, when, and why to implement activities, as well as advice about helping students avoid common pitfalls.

We hope that our approach to helping students learn physics will enhance your teaching experience, as it has our own and those of many colleagues who have used these materials in their classrooms.

14 Electric Charge: Force and Energy

14.1 | Qualitative Concept Building and Testing

14.1.1 Observe and find a pattern For the experiments that follow, you need two foam insulation tubes, a small piece of felt or wool, string, and plastic wrap. Suspend one tube from a string, as shown in the illustration. Before starting the experiments that follow, bring one end of tube 2 near one end of hanging tube 1. Is there any interaction? Now rub one end of each tube vigorously with different materials, as described below.

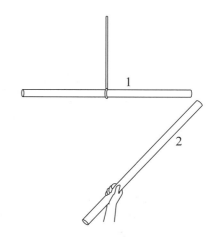

a. Bring the rubbed end of tube 2 or the material doing the rubbing near the rubbed end of the suspended tube 1 and record the behavior of tube 1.

Object 1	Object 2	Record your observation
Tube 1 rubbed with felt	Tube 2 rubbed with felt	
Tube 1 rubbed with felt	The felt that was used to rub tube 2	
Tube 1 rubbed with plastic wrap	Tube 2 rubbed with plastic wrap	
Tube 1 rubbed with plastic wrap	The wrap that was used to rub tube 2	
Tube 1 rubbed with plastic wrap	Tube 2 rubbed with felt	

b. Identify patterns in these observations. Devise an explanation.

14.1.2 Test your explanation Assemble two long pieces of nylon (stockings work fine) and a plastic grocery bag. Design an experiment to test how two pieces of nylon rubbed with the plastic bag interact with each other. Fill in the table that follows to make predictions about the outcomes of your experiments based on the patterns that you found in Activity 14.1.1.

Design an experiment to test how nylon pieces rubbed with plastic interact.	Predict the outcome of your experiment.	Perform the experiment and then describe the outcome.	Was the reasoning leading to your prediction correct? Explain.

14.1.3 Explain In Activities 14.1.1 and 14.1.2 you found a consistent pattern: Identical objects rubbed with a second material repel each other. The second material in turn attracts the objects it rubbed. Think of a mechanism that might explain why rubbing objects makes them attract or repel each other.

14.1.4 Test the explanation Your friend Gaurang says that electric interactions are the same as magnetic interactions because magnets also attract and repel each other. Consequently, he believes that when you rub objects, they become magnetized. Assemble a magnet on a swivel, two foam tubes, felt, and plastic wrap. Design an experiment whose outcome will allow you to decide whether Gaurang is correct or whether rubbing objects makes them participate in a different type of interaction.

Describe an experimental setup to test the idea that rubbing causes materials to become magnetic (Gaurang's idea).	Predict the outcome of your experiment based on Gaurang's idea.	Perform the experiment and then describe the outcome.	Make a judgment about Gaurang's idea based on the outcome.

14.1.5 Observe and explain You have two foam tubes; one tube is suspended at the center from a string, and the other is free. Vigorously rub one end of each tube with felt. Slowly bring the rubbed end of the free tube closer and closer to the rubbed end of the hanging tube. Describe your observations. What can you infer about how the electric force depends on the separation of the objects?

14.1.6 Observe and explain

a. Perform the experiments described and fill in the table that follows.

Experiment	Rub one end of foam tube 2 with felt and bring it close to a hanging foam tube 1 that has not been rubbed. Repeat, but this time rub the other end of tube 2 with plastic wrap and bring it near the end of unrubbed tube 1.	Rub one end of foam tube 2 with felt and bring it close to the end of a hanging metal rod 1 that has not been rubbed. Repeat, but this time rub the other end of foam tube 2 with plastic wrap and bring it near the end of the hanging metal rod 1.
Record what you observed.	Tube 2 rubbed with felt: Tube 2 rubbed with plastic wrap:	Tube 2 rubbed with felt: Tube 2 rubbed with plastic wrap:

b. Devise an explanation involving a possible internal structure of the foam material that might explain why the rubbed tubes attract the unrubbed tubes. Draw a charge diagram for the inside of the unrubbed tube.

c. Devise an explanation involving a possible internal structure of the metal rod that might explain why the rubbed tubes attract it. Draw a charge diagram for the inside of the metal rod.

d. Devise an explanation for why the metal rod responds differently than the foam tube.

e. List everyday experiences that are consistent with these observations.

14.1.7 Test your ideas Hang a small piece of aluminum foil from a 30-cm-long piece of thread, which is tied at the top to a plastic or wooden rod (for example, a ruler). Use the ideas that you devised in Activity 14.1.6 to predict what happens when you bring the end of a foam tube rubbed with fur near the piece of foil. Then repeat the procedure using a foam "peanut" (standard packing material) hanging from the thread. Fill in the table that follows.

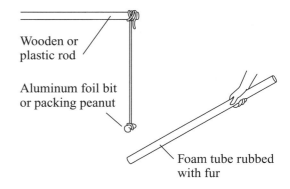

Wooden or plastic rod

Aluminum foil bit or packing peanut

Foam tube rubbed with fur

Experiment: Aluminum foil suspended on a string

Predict in words what you will observe.	Explain your prediction using a charge diagram.	Perform the experiment and record your observations.	Revise the explanation if necessary.

Experiment: Packing peanut suspended on a string

Predict in words what you will observe.	Explain your prediction using a charge diagram.	Perform the experiment and record your observations.	Revise the explanation if necessary.

14.1.8 Observe and explain An electroscope is made up of a vertical metal rod with a metal sphere top that sticks out of a glass enclosure; a cork-shaped piece of plastic prevents electric charge from going from the metal rod and metal leaves at the bottom onto the cylindrical metal enclosure. A thin metal leaf hangs down against the bottom of the rod inside the enclosure.

a. Perform the experiments described and fill in the table that follows.

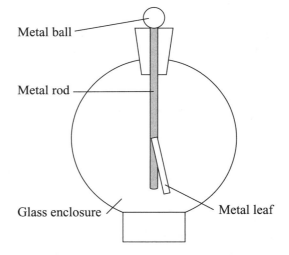

Metal ball

Metal rod

Glass enclosure

Metal leaf

Experiment	Record your observation.	Explain your observation.
Rub a foam tube with any material and then rub the top of the electroscope with the foam tube.		
Touch the top of the electroscope with your hand.		
Rub the top of the electroscope with a rubbed foam tube, remove the tube, and then touch the top of the electroscope with your hand while wearing a plastic glove.		

b. Modify the explanation you devised in Activity 14.1.7 to account for the outcomes of the third experiment.

14.1.9 Test your ideas Charge one electroscope by rubbing its top with a foam tube that has been rubbed with wool (the foam tube rubbed with wool is negatively charged). Place a second uncharged electroscope near the first charged electroscope.

a. Predict what happens if you touch a metal rod from the top of one electroscope to the top of the other. Wear a plastic glove to make sure you do not touch the rod with your bare skin. Explain your prediction, then perform the experiment and reconcile predictions with the outcome.

b. Discharge the electroscopes (touch them with your fingers) and predict what happens if you charge one electroscope and repeat the experiment in part a, only this time you touch a wooden rod from the top of the charged electroscope to the top of the uncharged electroscope. Fill in the table that follows to explain your predictions and then reconcile them with the observations.

Experiment: Touch the metal rod to the tops of the charged and uncharged electroscopes.

Predict in words what you will observe.	Explain your prediction.	Perform the experiment; record your observations.	Revise the explanation if necessary.

Experiment: Touch the wooden rod to the tops of the charged and uncharged electroscopes.

Predict in words what you will observe.	Explain your prediction.	Perform the experiment; record your observations.	Revise the explanation if necessary.

14.2 | Conceptual Reasoning

14.2.1 Represent and reason Several experiments are described below. Complete the table that follows.

Sketch the situation.	Describe the situation in words.	Draw a microscopic representation (charge diagram) of the charges inside the specified object.	Perform the experiment and describe the outcome.
	Rub a foam tube with fur and bring it near one end of an empty plastic bottle placed on a swivel.	Plastic bottle:	
	Rub a foam tube with fur and bring it *near* an electroscope without touching the electroscope.		

Sketch the situation.	Describe the situation in words.	Draw a microscopic representation (charge diagram) of the charges inside the specified object.	Perform the experiment and describe the outcome.
	Move the tube in the last experiment away from the electroscope.		
	A piece of aluminum foil rolled in a ball hangs vertically from a string near a charged foam tube.		The ball swings and touches the rubbed foam tube and then immediately swings back away from the tube.

14.2.2 Represent and reason Two positively charged objects are held near each other in the muzzle of a cannon (see part a of the illustration). When the "trigger" holding the cannonball is released, the positively charged cannonball flies out the end of the muzzle in part b. Certain types of energy have increased. Describe some type of energy decrease that you think might compensate for the increase in these other energies. *Note:* The situation shown in part a of the illustration is similar to that of a compressed spring; instead of the coils of the spring being squeezed together, two like charges are squeezed or pushed together. In part b of the illustration, this compressed electric "spring" is more relaxed. Fill in the table that follows.

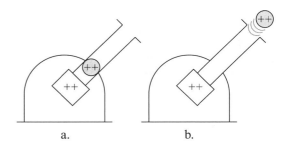

a.   b.

Using the language of energy, explain in words the described process, going from the initial state to the final state. Indicate your system choice.	Draw an energy bar chart representing the initial and final states.
	K_i U_{gi} U_{si} U_{qi} + W = K_f U_{gf} U_{sf} U_{qf} $\Delta U_{\delta U(\text{int})}$ + 0 — — — — — — — — — −

14.2.3 Represent and reason Imagine the energy changes of two opposite-sign charged objects used as a nutcracker, as illustrated in the figure to the right. What happens when the negatively charged block shown in a is released and moves near the nut, as shown in b? What type of energy decreases to make up for the increase in kinetic energy? Fill in the table that follows to answer these questions. *Note:* The situation shown in part a is similar to that of a stretched spring. Instead of the coils of a spring being stretched, the two opposite charges are pulled apart—like stretching a spring. In part b, this electric stretched "spring" is in the process of relaxing.

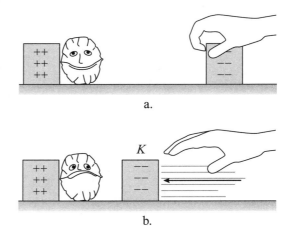

a.

b.

Using the language of energy, explain in words the process described above going from the initial state to the final state. Indicate your system choice.	Draw an energy bar chart representing the initial and final states.
	K_i U_{gi} U_{si} U_{qi} $+ W = K_f$ U_{gf} U_{sf} U_{qf} $\Delta U_{\delta U(int)}$

14.2.4 Reason

a. Does the analogy of a compressed spring for a system consisting of two similarly charged objects pushed close to each other make sense? Explain.

b. Does the analogy of a stretched spring for a system consisting of two oppositely charged objects pulled far apart make sense? Explain.

c. Discuss the limitations of both analogies.

14.3 | Quantitative Concept Building and Testing

14.3.1 Find a pattern Charles Coulomb used a torsion balance to measure the force that one charged ball exerts on another charged ball to find out how the force between two electrically charged objects depends on the magnitudes of the charges and on their separation. Coulomb could not measure the absolute magnitude of the electric charge on the metal balls. However, he could divide charges in half by touching a charged metal ball with an identical uncharged ball. The table that follows provides data that resemble what Coulomb might have collected. Find patterns in the data and devise a mathematical relationship based on these observations. Use graph paper to help. Remember to decide which are the independent variables and which is the dependent variable. Then analyze the changes in the dependent variable as you change only *one* independent variable at a time.

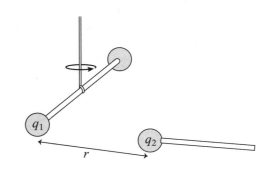

Charges (q_1, q_2)	Distance	Force
1, 1 (unit)	1 (unit)	1 (unit)
1/2, 1	1	1/2
1/4, 1	1	1/4
1, 1/2	1	1/2
1, 1/4	1	1/4
1/2, 1/2	1	1/4
1/4, 1/4	1	1/16
1, 1	2	1/4
1, 1	3	1/9
1, 1	4	1/16

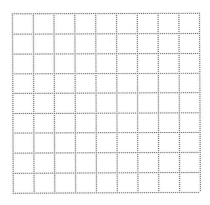

14.4 | Quantitative Reasoning

14.4.1 Represent and reason The metal balls on the cart in the illustration have equal magnitude charges and are very light. The rods supporting and connecting them are made of an insulating material and are also light. The cart rests on a smooth table.

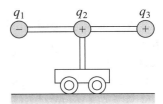

a. Fill in the table that follows. In this instance we consider only electric forces—not other types of force.

Draw labeled arrows representing electric forces exerted on the left metal ball. Represent the ball with a dot.	Draw labeled arrows representing electric forces exerted on the center metal ball.	Draw labeled arrows representing electric forces exerted on the right metal ball.	Draw labeled arrows representing electric forces exerted on the whole cart (a system with three charged balls).

b. Will the cart tend to accelerate either to the left or to the right? Explain your answer.

14.4.2 Represent and reason A positively charged ball of mass m hangs at the end of a string. Another positively charged ball is secured at the top end of the string to a wooden support. Fill in the table that follows.

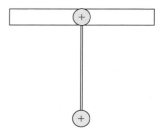

Draw a force diagram for the hanging ball if both balls are positively charged.	Represent the diagram mathematically using Newton's second law.	Draw a force diagram for the hanging ball if the top ball is negatively charged.	Represent the diagram mathematically using Newton's second law.

14.4.3 Represent and reason Two equal-mass stationary balls hang at the end of strings, as shown at the right. The ball on the left has electric charge $+5Q$, and the ball on the right has electric charge $+Q$. The strings make angles less than $45°$ with respect to the vertical plane. Fill in the table that follows.

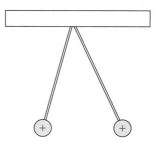

Draw a force diagram for the left ball.	Draw a force diagram for the right ball.	Decide which string makes a bigger angle with the vertical plane or if they make the same angle.
Apply Newton's second law in component form for the right ball. Horizontal x axis: Vertical y axis:	Based on your analysis, rank the forces $T_{S \text{ on } Q}$, $F_{5Q \text{ on } Q}$, and $F_{E \text{ on } Q}$, listing the largest force first. Explain the ranking:	

14.4.4 Equation Jeopardy The application of Newton's second law for a positively charged object at one instant of time is shown in the equation that follows. Other charged objects are along a horizontal line. Complete the table.

$$(9.0 \times 10^9 \, \text{N} \cdot \text{m}^2/\text{C}^2) \left[-\frac{(2.0 \times 10^{-4}\text{C})(3.0 \times 10^{-5}\text{C})}{(2.0 \, \text{m})^2} - \frac{(9.0 \times 10^{-4}\text{C})(3.0 \times 10^{-5}\text{C})}{(3.0 \, \text{m})^2} \right]$$

$$= (4.0 \, \text{kg})a_x$$

Draw a force diagram for the object at the instant the equation applies.	Sketch a situation the equation might describe at that particular instant.	Write in words a problem for which the equation is a solution (it applies at only one instant in time).	Determine one change that could be made in the situation so that the net force exerted on the object of interest is zero.

14.4.5 Evaluate the solution

The problem: A 2.0-kg cart with a $+2.0 \times 10^{-5}$ C charge on it sits at rest 1.0 m to the right of a fixed dome with charge $+1.0 \times 10^{-4}$ C. The cart is released. Determine how fast it is moving when it is 3.0 m from the fixed-charged dome.

Proposed solution: The situation is shown at the right.

Simplify and diagram

We assume that the dome and cart are point particles.

See the force diagram to the right.

Represent mathematically and solve

$\Sigma F_x = kq_1q_2/r = ma_x$

$a_x = kq_1q_2/rm$

$\quad = (9 \times 10^9 \, \mathrm{N \cdot m^2/C^2})(1.0 \times 10^{-4} \, \mathrm{C})(2.0 \times 10^{-5} \, \mathrm{C})/(1.0 \, \mathrm{m})(2.0 \, \mathrm{kg}) = 9.0 \, \mathrm{m/s^2}$

$v^2 = 0^2 + 2(9.0 \, \mathrm{m/s^2})[\,(3.0 \, \mathrm{m}) - (1.0 \, \mathrm{m})\,] \quad$ or $\quad v = 18 \, \mathrm{m/s}$

a. Identify any missing elements or errors in the solution.

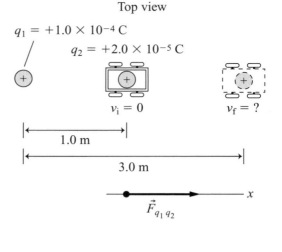

Top view

$q_1 = +1.0 \times 10^{-4} \, \mathrm{C}$

$q_2 = +2.0 \times 10^{-5} \, \mathrm{C}$

$v_i = 0 \qquad v_f = ?$

1.0 m

3.0 m

$\vec{F}_{q_1 q_2}$

x

b. If there are errors, provide a corrected solution or the missing elements.

14.4.6 Represent and reason A negatively charged ball, initially at rest, falls until it hits a massless spring, which it compresses while stopping. The bottom of the spring rests on a positively charged block.

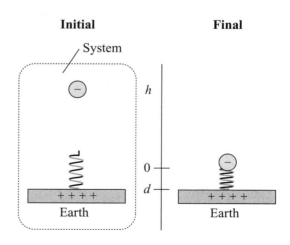

Initial

Final

System

h

0

d

Earth

Earth

Draw a bar chart consistent with the process.	Apply the generalized work–energy equation to the process.
$K_i \; U_{gi} \; U_{si} \; U_{qi} + W = K_f \; U_{gf} \; U_{sf} \; U_{qf} \; \Delta U_{\delta U(int)}$ + 0 — — — — — — — −	

14.4.7 Represent and reason

a. Chris releases the trigger on an electric cannon. The cannonball with charge $+q$ and mass m fires vertically upward due to its repulsion from the stationary ball with a charge $+Q$. The cannonball reaches the apex of its flight at distance h above its starting position. Represent the process physically with a bar chart and mathematically in part a of the table that follows.

b. Now, suppose that the charge $+Q$ is reduced to $+Q/2$. Represent this process with a bar chart and mathematically. Describe in words how reducing the charge affects the process in part b of the table.

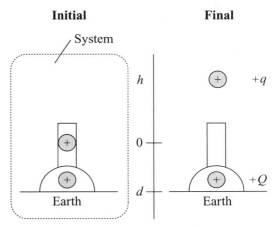

Draw a bar chart consistent with the process.	Apply the generalized work–energy equation.
a. $K_i \; U_{gi} \; U_{si} \; U_{qi} + W = K_f \; U_{gf} \; U_{sf} \; U_{qf} \; \Delta U_{\delta U(int)}$ + 0 — — — — — — — −	
b. $K_i \; U_{gi} \; U_{si} \; U_{qi} + W = K_f \; U_{gf} \; U_{sf} \; U_{qf} \; \Delta U_{\delta U(int)}$ + 0 — — — — — — — −	

14.4.8 Bar–chart Jeopardy The bar charts below could represent many processes. Complete the table for each bar chart.

Bar charts for unknown processes	$K_i \quad U_{gi} \quad U_{si} \quad U_{qi} \; +W= \; K_f \quad U_{gf} \quad U_{sf} \quad U_{qf} \quad \Delta U_{\delta U(int)}$	$K_i \quad U_{gi} \quad U_{si} \quad U_{qi} \; +W= \; K_f \quad U_{gf} \quad U_{sf} \quad U_{qf} \quad \Delta U_{\delta U(int)}$
Draw an initial–final sketch of one possible process described by the bar chart.		
Describe the process in words.		
Convert the bar chart into the work–energy relationship as applied to this process.		

14.4.9 Regular problem Consider a simplified version of a real situation that occurs in the center of our sun. A proton of charge $+e$ moves directly toward a stationary deuterium nucleus, also of charge $+e$. (We assume that the deuterium does not move.)

a. Determine the speed that the proton must move toward the deuterium when far from it so that it is able to get within 1.0×10^{-15} m before stopping. (At this distance, there is a good chance that the proton and deuterium will fuse to form a helium nucleus. The fusion releases considerable energy.)

Sketch and translate	Simplify and diagram (construct a bar chart)
Sketch the initial and final problem states. Choose a system.	
Represent mathematically	**Solve and evaluate**
Use the bar chart to apply a generalized work–energy equation.	

b. Estimate the temperature of hydrogen when this fusion can occur with good probability. The proton mass is 1.67×10^{-27} kg. What assumptions did you make?

14.4.10 Equation Jeopardy The equation below describes one or more physical processes.

$$\frac{1}{2}\left(1.67 \times 10^{-27}\,\text{kg}\right)v_{\text{i}}^2 + \frac{1}{2}\left(1.67 \times 10^{-27}\,\text{kg}\right)v_{\text{i}}^2 = \frac{\left(9.0 \times 10^9\,\text{N}\cdot\text{m}^2/\text{C}^2\right)\left(1.6 \times 10^{-19}\,\text{C}\right)^2}{1.0 \times 10^{-15}\,\text{m}}$$

Draw a bar chart that is consistent with the equation.	Sketch the initial–final states that the equation might describe.	Write in words a problem for which the equation could be a solution.

14.4.11 Evaluate the solution

The problem: A 2.0-kg cart with a $+2.0 \times 10^{-5}$ C charge on it starts at rest 1.0 m from a fixed dome with charge $+1.0 \times 10^{-4}$ C. The cart is released. Determine how fast it is moving when it is 3.0 m from the fixed-charged dome.

Proposed solution:

Sketch and translate

The situation is depicted in the illustration to the right.

Simplify and diagram

We assume that the dome and cart are point particles.

   See the bar chart to the right.

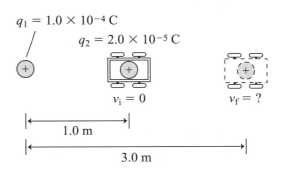

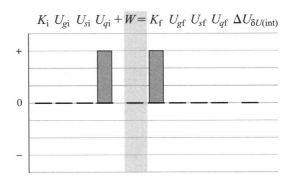

Represent mathematically and solve

$$U_{qi} = K_f$$
$$kq_1q_2/r = (1/2)mv^2$$
$$v = [2kq_1q_2/rm]^{1/2}$$
$$= [2(9 \times 10^9)(1 \times 10^{-4})(2 \times 10^{-5})/(1.0)(2.0)]^{1/2}$$
$$= 18 \text{ m/s}$$

a. Identify any missing elements or errors in the solution to the problem.

b. If there are errors, provide a corrected solution or missing elements.

14.4.12 Regular problem One simple and productive model of a hydrogen atom (although rarely used in modern physics) is a positive nucleus (a proton) and a negatively charged electron moving around it in a circular orbit. Estimate the electron's speed in this model. The radius of the atom is 0.51×10^{-10} m.

14.4.13 Regular problem Use the result of Activity 14.4.12 to determine the minimum energy that the proton nucleus–electron system needs to gain for the electron to become free (to remove the electron far from its proton nucleus).

14.4.14 Design an experiment Your group is working on a static electricity project. You need to use a nonconducting string from which you will hang pieces of aluminum foil. Your friend brings two kinds of dental floss. Design an experiment to find out which floss is conducting and which one is not. Describe the experiment and explain how you will make a decision based on its outcome.

15 Electric Fields

15.1 | Qualitative Concept Building and Testing

15.1.1 Represent and reason Imagine two pointlike charged objects of mass m_1 and m_2 that have electric charges q_1 and q_2, respectively. Complete the table that follows and analyze the objects' gravitational and electrostatic interactions using the filled cells as hints.

Problem	Gravitational interaction	Electrostatic interaction
What property of objects determines whether they participate in the interaction?		Electric charge
What is the direction of the force between the interacting objects?	It is an attractive force.	
Write an expression for the magnitude of the force between interacting objects.		
How does the magnitude of the force depend on properties of the objects?	It is directly proportional to m_1 and directly proportional to m_2.	
How does the magnitude of the force depend on the distance between the objects?		
Write an expression for the potential energy of the interacting objects.	$U_g = -G\dfrac{m_1 m_2}{r}$	$U_e = k\dfrac{q_1 q_2}{r}$

15.1.2 Reason Discuss similarities and differences between gravitational and electrostatic interactions. Suggest possible mechanisms for how these interactions can occur at a distance—without direct contact between objects.

15.1.3 Observe and explain Assemble an uncharged electroscope; a foam tube; fur; a small, clear, glass beaker (or plastic cup); and a small metal cup. Perform the experiments described in the table below and explain the outcomes using the idea of an electric field.

Experiment	Result	Sketch the situation.	Explain the experiment using the idea of an electric field.
Bring the charged end of a foam tube down toward the top of an uncharged electroscope without touching the electroscope.	The needle of the electroscope deflects.		
Take the charged end of the foam rod away from the top of the uncharged electroscope.	The needle of the electroscope goes back to its original position.		
Repeat the first two experiments, this time using a small glass beaker to cover the metal ball on the electroscope.	The needle behaves the same way in both cases, but the deflection is a little smaller.		
Repeat the first two experiments, only this time using a small metal cup to cover the metal ball on the electroscope.	There is no deflection of the needle, as if the charged tube was not brought close.		

15.1.4 Design an experiment Use any equipment you have to design an experiment that allows you to observe that the electric interaction between two objects can be blocked or shielded. Perform the experiments and record the outcomes.

15.2 | Conceptual Reasoning

15.2.1 Represent and reason

a. For each situation pictured in the table that follows, represent at the dots the gravitational force or the electric force that the source mass or source charge exerts on a test mass or test charge. Fill in the table that follows.

Word description	Represent with arrows the gravitational force that the Earth (the source mass) exerts on small objects (called *test masses*) at the points shown. Draw the arrows with the correct relative lengths.	Represent with arrows the electric force that the object with a large negative charge (the source charge) exerts on small positively charged objects (called *test charges*) at the points shown.	Represent with arrows the electric force that the object with a large positive charge (the source charge) exerts on small positively charged objects (called *test charges*) at the points shown.
Picture description			

b. Use a field approach to explain in words how the source object can exert a force on test objects without directly touching them at each of these points. Discuss how the magnitude of this force might depend on the magnitude of the gravitational field, on the magnitude of the mass of the test objects in the gravitational field, on the magnitude of the electric field, and on the test charges in the electric field.

c. Discuss how the presence of a source mass or a source charge alters the space. How far do you think this alteration extends?

d. Discuss whether a system that includes Earth and a test object possesses gravitational potential energy. Discuss whether a system that includes the source-charged object and the test-charged object possesses electric potential energy. Does each energy depend on the magnitude of the source mass or the test mass? On the source charge or the test charge?

15.2.2 Represent and reason Estimate the direction and the magnitude of the $\vec{E}$ field at points *A*, *B*, and *C* in the figure that follows.

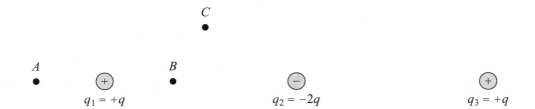

15.2.3 Represent and reason Estimate, by drawing on the figures that follow, the direction and relative magnitude of the $\vec{E}$ field at points *A*, *B*, and *C* due to the electric dipole on the heart (shown at one instant during a heartbeat cycle).

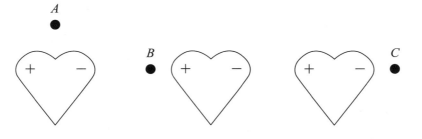

15.2.4 Reason Use the analogy between the electric field and the gravitational field to estimate the Earth's gravitational field $\vec{g}$ at the points shown in the figure.

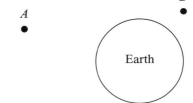

a. What would you choose as the source-mass object? What would you use as the test-mass object?

b. Draw $\vec{g}$ field vectors at points *A, B,* and *C*.

c. Discuss how the magnitude and direction of the $\vec{g}$ field are related to the acceleration of free-falling objects placed at these points.

15.2.5 Diagram Jeopardy $\vec{E}$ field vectors due to one or more electrically charged objects are shown below. Indicate with circles, including + or − signs, the locations of the charged objects causing the fields.

a.

b.

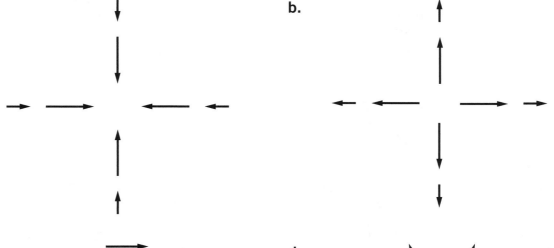

c.

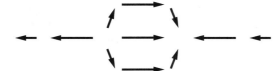

d.

15.2.6 Reason Answer the following questions:

a. Can electric field lines cross? Explain.

b. What is the direction of the $\vec{E}$ field line at a point midway between two equal-magnitude, positively charged objects? How do you know?

c. What is the direction of the $\vec{E}$ field line at a point midway between two equal-magnitude, oppositely charged objects? How do you know?

15.2.7 Represent and reason Draw $\vec{E}$ field lines for the electric field created by the source-charged objects described in the table that follows. Show the vectors that are tangent to the lines.

a. A pointlike positively charged object	b. A pointlike positively charged object with twice the magnitude of charge as in part a	c. A pointlike negatively charged object	d. A pointlike negatively charged object with twice the magnitude of charge as in part c

e. Two positively charged pointlike objects of equal-magnitude charge, separated by a distance *s*	f. Two negatively charged pointlike objects of equal-magnitude charge, separated by a distance *s*	g. A small positively charged object and a small negatively charged object of equal-magnitude charge, separated by a distance *s*	h. A small positively changed object and a small negatively charged object with twice the magnitude of electric charge, separated by a distance *s*

15.2.8 Represent and reason Imagine that a small, positively charged object moving toward the top of the page enters an electric field with the lines shown below.

a. Sketch on the illustration an approximate path of the object as it moves through the field. The direction of the initial velocity of the object as it enters the field is shown in the figure.

b. Discuss whether the lines represent the paths that a charged object follows after it enters the field.

15.2.9 Represent and reason The figure on the right shows the instant when a hollow metal box is placed in a uniform electric field (its effect on the external field is not shown).

a. Indicate the electric charge distribution in the metal due to the external electric field. *Note:* If electrons move from one part of the box to another, the part with a deficiency of negatively charged electrons is now positively charged, and the part with an excess of electrons is negatively charged.

b. Draw electric field lines caused by this induced-charge distribution on the surface of the box and discuss the magnitude of the total $\vec{E}$ field inside the box.

c. Discuss how your reasoning in parts a and b helps explain why it is safe to sit in a car during a lightning storm.

d. Draw the new shape of the field lines outside the box. How does the redistribution of electrons inside the box affect the electric field outside?

15.2.10 Reason Complete the table that follows.

Compare the electric potentials at points A, B, and C due to the positive source charge, the largest potential listed first. Show at least one other point at which electric potential is the same as at point A.	Compare the electric potentials at points D, E, and F, the largest potential listed first. (*Note:* A large negative number is less than a small negative number.) Show at least one other point at which electric potential is the same as at point E.
• C • B • A (+)	• F • E • D (−)

15.2.11 Represent and reason In the table that follows, draw lines of equal gravitational potential caused by Earth, a mass source for the gravitational potential. Think of how you can write an expression for the gravitational potential that is analogous to the expression for the electric potential.

Far away from the surface of the Earth, when the Earth is modeled as a sphere	Close to the surface of the Earth, when the surface of the Earth can be modeled as a plane
Earth	_____ Earth

15.2.12 Represent and reason Using the same source charges as in Activity 15.2.7 a–d and the electric field lines that you drew there, draw the surfaces of equal potential. Find a pattern between the direction of the lines and the change of electric potential (whether the electric field lines point in the direction of increasing or decreasing potential).

15.2.13 Explain Sometimes physicists use the analogy between lines of equal electric potential and lines of equal altitude on topographical maps.

a. Explain how this analogy works and why it is useful.

b. Describe how the closeness of the equal-altitude lines relates to the steepness of a mountain.

c. Use this analogy to draw electric field lines for the electric field whose equal potential surfaces look as follows. Explain.

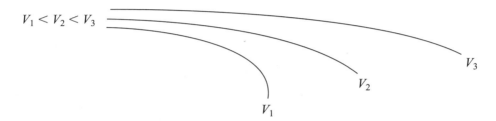

$$V_1 < V_2 < V_3$$

V_3

V_2

V_1

15.2.14 Reason

a. Draw equal potential surfaces and $\vec{E}$ field lines for a negatively charged, infinitely large metal plate. For help, complete the activities in the table that follows.

Sketch the plate with electric charges (shown below).	Draw lines of equal electric potential. Indicate where the potential is higher.	Draw $\vec{E}$ field lines.

b. Why are the $\vec{E}$ field lines perpendicular to the surface of the plate and to the equal potential surfaces? Explain.

c. Are both the $\vec{E}$ field lines and equal potential surfaces equally spaced? Explain.

d. Examine the locations of the $\vec{E}$ field lines and the equal potential surfaces with respect to each other. Discuss any patterns that you find. Explain.

15.3 ▪ Quantitative Concept Building and Testing

15.3.1 Derive A pointlike object of charge q (a source charge) is located at the origin of a coordinate system.

a. Use the definition of $\vec{E}$ field to derive a relationship for the magnitude of the $\vec{E}$ field caused by this charged object at some point A a distance r from the charged object. Represent the relationship graphically (the magnitude of $\vec{E}$ -versus-r).

b. Use the definition of electric potential to write an expression for the electric potential at point A due to the pointlike charge q at the origin. Point A is a distance r from the charge q. Represent the relationship graphically (V-versus-r).

c. Discuss whether there is any relationship between the $\vec{E}$ field and electric potential.

15.3.2 Reason

a. Imagine that you have a positively charged, solid metal ball. Complete the table that follows.

Indicate in the drawing how the charge is distributed inside the ball and on the ball's surface and explain your drawing.		Draw electric field lines outside the ball and compare their distribution with the distribution of electric field lines of a pointlike charge.	
Draw electric field lines inside the ball (if any). Explain.		Draw a graph of the magnitude of the $\vec{E}$ field versus the distance r from the center of the ball.	
		Draw a graph of the V field versus the distance r from the center of the ball.	

b. Discuss how you can have a situation in which the electric field at some location is zero but the electric potential is not. Does this seem reasonable? Explain.

15.3.3 Reason You have two metal spheres of radii R_1 and R_2. The sphere on the left has a charge $+q_1$, and the sphere on the right is not charged.

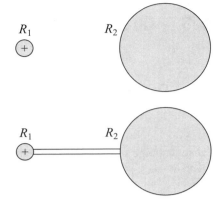

a. Explain whether the electric potential on the surfaces of the spheres will be the same or different before they are connected with a metal rod.

b. Will the electric potential of the surfaces be the same or different after they are connected?

c. Which sphere will have a larger charge after the connection? How does the total charge of the two spheres after the connection compare to the initial charge of the left sphere?

d. Discuss how the situation in parts a–c is related to the method of discharging objects by connecting them to Earth—so-called *grounding*. Earth is a huge conductor, like a huge metal sphere.

15.4 | Quantitative Reasoning

15.4.1 Represent and reason A small aluminum ball is charged with $+1.0 \times 10^{-9}$ C. Determine the magnitude of the $\vec{E}$ field due to this source charge at the four points shown below. Represent the $\vec{E}$ field at each point using an arrow.

5 cm to the right of the ball	7 cm to the left of the ball	1 cm above the ball	10 cm below the ball

©2014 Pearson Education.

15.4.2 Represent and reason Charged objects $q_1 = +q$ and $q_2 = -q$ are separated by a distance r, as shown in the art below. Determine an expression for the magnitude of the $\vec{E}$ field at points $(r, 0)$ and $(0, r)$ due to these two charges. Fill in the table that follows.

Sketch	Draw an electric field diagram.	Determine the component fields E_x and E_y.	Determine the magnitude of the net electric field.

15.4.3 Represent and reason A 0.060-kg electrically charged ball hangs at the end of a string oriented 53° outward from a charged vertical plate. The $\vec{E}$ field produced by the plate at the position of the ball is 1.0×10^5 N/C and points away from the plate. Complete the table that follows to determine the charge on the ball (sign and magnitude). Assume that the gravitational constant is 10 N/kg.

Sketch the situation.	Draw a force diagram for the hanging ball.	Represent the diagram mathematically using Newton's second law.	Determine the charge on the ball.
		x: y:	

15.4.4 Equation Jeopardy An electrically charged block moves on a horizontal surface. The application of Newton's second law and kinematics to this situation is shown in the equations that follow. Complete the table. Assume that the gravitational constant is 10 N/kg. *Note: n* is the magnitude of the normal force, and N is the unit of force. The zeros are for forces that have zero components along that axis.

$$\Sigma F_x = (+1.0 \times 10^{-5}\,\text{C})(+2.0 \times 10^6\,\text{N/C}) + 0 - 0.40\,n + 0 = (4.0\,\text{kg})a_x$$

$$\Sigma F_y = 0 + n + 0 - (4.0\,\text{kg})(10\,\text{N/kg}) = 0$$

$$2a_x(24\,\text{m} - 8.0\,\text{m}) = v^2 - 0$$

Draw a force diagram for the block.	Sketch a situation the equation might describe.	Write in words a problem for which the equation is a solution.	Determine one change that could be made in the situation so that the net force exerted on the object of interest is zero.

15.4.5 Evaluate the solution

The problem: A 0.040-kg cart is moving at a speed of 6.0 m/s when it enters a 1.8×10^4 N/C electric field that stops the cart in 0.40 m. Determine the electric charge on the cart.

Proposed solution:

Sketch and translate

The situation is shown in the illustration.

Simplify and diagram

We assume that the cart and its charge are a point-like object and that there is no friction.

A force diagram for the cart is shown at the right.

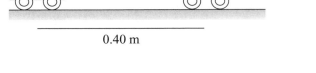

0.40 m

Represent mathematically, solve and evaluate

$$a = (v_f^2 - v_i^2)/2(x_f - x_i) = (6.0\,\text{m/s})^2/2(0.40\,\text{m}) = 90\,\text{m/s}^2$$

$$qE = ma$$

$$q = ma/E = (0.040\,\text{kg})(90\,\text{m/s}^2)/(1.8 \times 10^4\,\text{N/C}) = 2.0 \times 10^{-4}\,\text{C}$$

a. Identify any errors or missing elements in the solution to this problem.

b. Provide a corrected solution if there are errors.

15.4.6 Regular problem A 0.50-kg cart with charge $+3.0 \times 10^{-6}$ C slides on a horizontal surface that exerts a 2.0-N friction force on the cart. The cart moves in a 1.0×10^{-6} N/C horizontal constant $\vec{E}$ field that points toward the right. If the cart starts at rest, determine its speed after moving 2.0 m.

Sketch the situation.	Draw a force diagram for the cart.	Represent the situation mathematically.	Determine the cart's speed.

15.4.7 Represent and reason The same situation is represented in different ways in the table.

Words
Determine the electric potential V at a distance r above a point directly between an electric dipole, which consists of an object of charge $+Q$ on the left and a second object of charge $-Q$ that is a distance r to the right.

Sketch

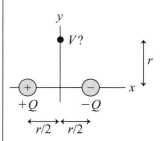

Mathematical representation and solution

$$V = \frac{k(+Q)}{\left[\left(\frac{r}{2}\right)^2 + r^2\right]^{\frac{1}{2}}} + \frac{k(-Q)^2}{\left[\left(\frac{r}{2}\right)^2 + r^2\right]^{\frac{1}{2}}} = 0$$

Are the representations consistent with each other? Explain.

15.4.8 Represent and reason Determine the speed of the dust particle illustrated in the table that follows.

Words	Sketch
A 0.0010-g dust particle charged to $+1.0 \times 10^{-7}$ C encounters a uniform 1.0×10^4 N/C electric field, as shown. Determine the change in its speed after it moves a vertical distance of 1.0 m.	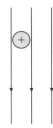
Represent mathematically and solve.	
Indicate any assumptions you made.	

15.4.9 Represent and reason Determine the electric potential at point A. Specify your choice of the reference point where the potential is zero.

Words
Three charged objects are shown. Determine the electric potential V at point A—that is, at position $x = 0.9$ m.

Sketch

$q_1 = +2 \times 10^{-6}$ C $q_2 = -4 \times 10^{-6}$ C $q_3 = +2 \times 10^{-6}$ C

$x_1 = 0$ $x_2 = 0.3$ m $x_3 = 0.6$ m $x_A = 0.9$ m

Represent mathematically and solve.

15.4.10 Evaluate the solution

The problem: Imagine two electrically charged objects $+Q$ and $-Q$ connected by a plastic rod (not shown in illustration) of length l. Write an expression for the electric potential at position I (distance r to the left of $+Q$), position II (halfway between the charged objects), and position III (distance r to the right of $-Q$). Identify and correct any errors in the solutions in the table that follows.

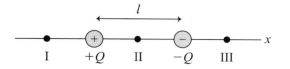

Proposed solution	Identify errors in the solution (if there are any).	Provide a corrected solution (if there are errors).
$V_I = kQ/r^2 + k(-Q)/(2r)^2$ $\quad = 3kQ/4r^2$		
$V_{II} = kQ/(r/2) + k(-Q)/(r/2)$ $\quad = 0$		
$V_{III} = kQ/2r + kQ/r$ $\quad = 3kQ/2r$		

15.4.11 Represent and reason Complete the table that follows. Do not solve for anything.

Description in words of the process	Rocket launch: A 0.010-kg rocket with a $+1.0 \times 10^{-3}$ C charge rests on a very large horizontal plate. Another very large horizontal plate above the first has a hole in it directly above the rocket. When a switch is closed, the lower plate and upper plates become charged so that there is a +10,000-V electric potential on the lower plate relative to zero potential on the upper plate. The rocket shoots up through the hole. Determine the maximum height that the rocket reaches above its starting position. (*Note:* The electric potential does not change after the rocket passes through the hole in the top plate.)
Sketch the process.	

(continued)

Construct a qualitative bar chart.	K_i U_{gi} U_{si} U_{qi} $+ W = K_f$ U_{gf} U_{sf} U_{qf} $\Delta U_{\delta U(\text{int})}$ (blank bar chart with +, 0, − axis)
Write a mathematical description of the process.	

15.4.12 Equation Jeopardy
The generalized work–energy equation is applied to a physical process in the last row of the table that follows. Complete the table for this process.

Write a word description of the process.	
Sketch the process.	
Construct a qualitative bar chart.	K_i U_{gi} U_{si} U_{qf} $+ W = K_f$ U_{gf} U_{sf} U_{qf} $\Delta U_{\delta U(\text{int})}$ (blank bar chart with +, 0, − axis)
Mathematical description of the process	$(2.0 \times 10^{-5}\,\text{C})(40{,}000\,\text{V})$ $= (1/2)(1.0 \times 10^{-3}\,\text{kg})v^2$ $+ (1.0 \times 10^{-3}\,\text{kg})(10\,\text{N/kg})(40\,\text{m})$

15.4.13 Regular problem To cleanse the air of dust and pollen, some homes have electrostatic precipitators in their heating and air-conditioning systems. These units work by moving particle-laden air through an ionizing area, in which the particles acquire a charge, and then into an area in which an electric field is present. Because the particles are charged, they experience a force and are attracted to a collector that home owners need to clean from time to time. Such units are also used on a larger scale as industrial scrubbers in smokestacks, where particle-laden air rises through the stack, acquires a charge by an industrial ionizing source, and is filtered via electrostatic attraction. Imagine that you are a member of a team from the Environmental Protection Agency, which is trying to determine whether the school's heating smokestack is effective in removing most particulate matter. You find that the smoke particles move up the 20-m-high stack at a constant 5-m/s speed. By checking the ionizing equipment, you deduce that the specific charge (charge per unit mass) imparted to each particle is 1×10^{-5} C/kg. Most particles have a mass of $1 \mu g$ (10^{-6} g), but some are as large as $100 \mu g$. You see that two plates separated by 0.30 m are on opposite sides of the chimney, with a 3000-V/m $\vec{E}$ field between them. Will you recommend that the operating license for the smokestack be renewed? Support your ruling with a careful analysis.

16 DC Circuits

16.1 | Qualitative Concept Building and Testing

16.1.1 Observe and explain You have two electroscopes, initially uncharged, situated near each other. Explain the outcomes of each experiment described in the table that follows.

Experiment	Rub a foam tube with wool and then rub the top of one electroscope with the rubbed part of the tube. Observe that the needle of the electroscope deflects.	Then while wearing latex gloves connect a metal wire between the top of the first electroscope and the top of the second uncharged electroscope. The leaves of the second electroscope instantly separate, and the leaves of the first electroscope come closer to each other but do not go down completely.	Now rub the first electroscope with the charged tube again. The leaves of both electro-scopes deflect more.
Explain using the language of *V* field (electric potential).			

16.1.2 Observe and explain You again have two electroscopes situated near each other; one is charged, and the other is not. This time the leads of a neon lightbulb connect the electroscopes (you may need an extra wire connected to one of the bulb's leads). Fill in the table that follows.

Experiment	Use the bulb to connect a charged electroscope and an uncharged electroscope. You observe a short flash of light when you first connect them.	Then charge both electro-scopes and touch the neon bulb leads across the tops of the electroscopes. You observe no light.
Explain using the language of *V* field (electric potential).	Why does the flash last just a short time interval?	

16.1.3 Observe and explain You have a Wimshurst generator and a neon lightbulb. Complete the table that follows.

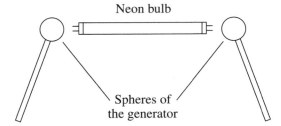

Neon bulb

Spheres of
the generator

Experiment	Crank the handle of the Wimshurst generator. When you bring the metal spheres of the generator near each other, there is a strong spark.	Crank the handle of the generator with the metal spheres about 5-cm apart. Touch the lead wires of a neon lightbulb to the metal spheres of the generator. You see a bright flash of light from the neon bulb.
Explain		

16.1.4 Observe and explain You charge a Wimshurst generator by cranking the handle and then hang a Ping-Pong® ball with a metal coating from a thin thread between the charged spheres of the generator. Fill in the table that follows.

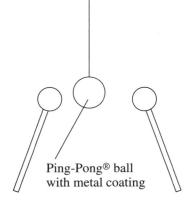

Ping-Pong® ball
with metal coating

Ping-Pong® is a registered trademark of Parker Brothers, Inc.

Experiment	The metal-coated ball swings rapidly back and forth between the oppositely charged spheres of the generator for about 30 s but eventually slows down and stops.
Explain using the language of *V* field and energy	

16.1.5 Observe and explain You have a battery, a wire, and a lightbulb. Try different arrangements of these three elements to make the lightbulb glow.

Draw pictures of the arrangements where the bulb lights and several where it does not. Explain how this experiment is similar to 16.1.3 and 16.1.4 and how it is different.

16.1.6 Explain Use your observations and explanations for the previous activities to answer the following questions.

a. Summarize the conditions that are necessary for the continuous flow of electric charge through a metal wire.

b. What properties of the device connected to the wire are necessary to maintain a continuous electric charge flow through the wire? Think of such a device in your everyday experience.

16.1.7 Test your ideas One of the devices that maintains a continuous potential difference across different points of a circuit is a battery. Use a 45-V battery for the following testing experiment.

a. Fill in the table that follows.

Experiment	Predict the outcome.	Perform the experiment and record the outcome. Explain.
What happens if you connect the poles of the battery to parallel metal plates placed near each other with a metal-coated Ping-Pong® ball hanging from a thin thread between the plates?		

b. Explain ways in which this experiment is analogous to a battery connected to a lightbulb (see Activity 16.1.5). Indicate the corresponding parts of the two different systems described in the two different activities. How are the systems different?

16.1.8 Observe and design Draw circuit diagrams according to the word descriptions below. Build the circuits, and observe the relative brightness of the lightbulbs.

Circuit description	Draw a circuit diagram.	Discuss the brightness of the lightbulb (is it more bright or less bright than in the first experiment).
One 1.5-V battery, one lighted light-bulb, and wires		
Two 1.5-V batteries arranged so that the positive side of one touches the negative side of the other, forming a chain (in physics they are said to be *in series*); one lighted lightbulb; and wires		
Two 1.5-V batteries arranged so that their positive sides are together and negative sides are together, forming a ladder (in parallel); one lighted light-bulb; and wires		

16.2 | Conceptual Reasoning

16.2.1 Reason You learned in Section 16.1 that for a lightbulb to glow, the two poles of a battery must be connected to the lightbulb with conducting wires. You also observed experiments in which several batteries were connected to lightbulbs in series or in parallel. Use your knowledge of the internal structure of conductors and the understanding of the role of a battery to explain these observations using two analogies: one involving flowing water and the other involving a group of people running on a track. Remember that an analogy does not need to account for all aspects of a phenomenon. However, if you find similar aspects, make a note of them.

Parts of the electric circuit	Parts of the water system	Parts of the running people system
Moving electrons		
Battery		
Connecting wires	Pipes with water in them	
Lightbulb		Muddy patch on the track

Observed properties of the electric current	Observed properties of the water system	Observed properties of the running people system
When batteries are in series, the lightbulb is brighter.		
When identical batteries are in parallel, the lightbulb is the same brightness.		

16.2.2 Reason In Chapter 15 you learned the concept of V field or potential difference. Use this concept to explain the role of a battery in a circuit.

a. Describe an analogy between some part of a system with water flow in a pipe caused by a pump and the potential difference provided by a battery in a circuit.

b. Describe an analogy between some part of a system with water flow in a pipe caused by a pump and the physical quantity *electric current* in a circuit.

16.2.3 Reason Use the flowing-water system and running-people system to find analogies for the quantities *potential difference* and *electric current*. Fill in the table that follows.

Electric circuit	Water system	Running people
Potential difference between two points ΔV		
Current I through a wire		

16.2.4 Observe and explain

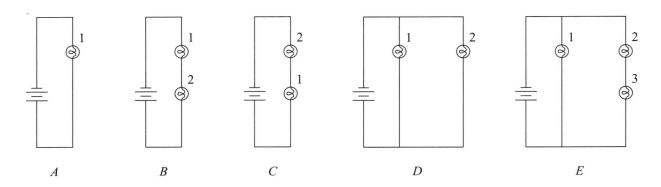

A   B   C   D   E

a. Build circuit *A* using a battery, wires, and identical bulbs (see the figures above) and then notice the brightness of the bulb. Then build circuit *B* and notice the brightness of the bulbs. Explain the differences in your observations using the concept of *V* field (potential difference) or any of the analogies.

b. Build circuit *C* and notice the brightness of the bulbs. Explain your observations using the concept of current and any analogies.

c. Build circuit *D* and notice the brightness of the bulbs. Now build circuit *E* and notice the brightness of the bulbs. Explain the differences in your observations using the concepts of potential difference and current.

d. Can you say that a battery is a source of constant current? Explain your answer.

16.2.5 Reason Use the analogies you discussed in Activity 16.2.2 and the ideas of potential difference and current to rate the bulbs in the circuit shown at the right according to their brightness, listing the brightest bulb first. Indicate whether any bulbs are equally bright. Explain your ratings.

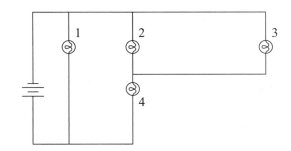

16.2.6 Reason

a. Rate the bulbs in the circuit shown to the right according to their brightness when the switch is open.

b. Now rate the bulbs in the circuit when the switch is closed.

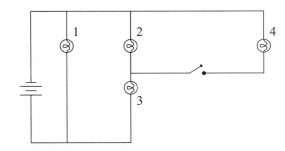

c. Indicate how the brightness of the first three lightbulbs changes after the switch is closed.

16.2.7 Represent and reason

a. Predict how the brightness of the top bulb shown in the illustration to the right changes when you close switch 1.

b. Predict how the brightness of the top bulb changes when you close switch 2 (switch 1 is open).

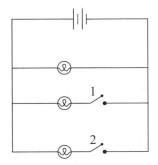

c. Rate the brightness of the three bulbs when both of the switches are closed. Explain your ratings.

d. Perform the experiment, record your observations, and compare them to your predictions. Explain the discrepancies

16.3 | Quantitative Concept Building and Testing

16.3.1 Observe, find a pattern, and explain You connect a resistor in series with an ammeter and a variable potential difference source. In the table, the electric current I through the resistor is shown as you vary the potential difference ΔV across the resistor.

Potential difference ΔV (V)	0.0	2.0	4.0	6.0	8.0	10.0
Current I (A)	0.000	0.020	0.038	0.061	0.079	0.105

Complete the table that follows.

Construct a circuit diagram.	Represent data graphically.	I (A) ΔV (V)
Describe the pattern mathematically.	Explain the relationship between the current through the resistor and the potential difference across it.	

16.3.2 Test your idea Design an experiment to test whether the linear relationship for current and potential difference holds for different resistors. Use commercial resistors and lightbulbs.

Describe experiments in words and specify what type of resistive device you will study.	Draw an electric circuit and measure the value of the resistance of the resistor using an ohmmeter (an electric device that measures the electric resistance of an object).	Write your prediction using the relationship $I = (1/R)\Delta V.$	Perform the experiment, record the outcome, and decide whether the relationship holds for this particular type of resistor.

16.3.3 Observe and explain The readings of three ammeters are shown at the right. What can you say about the magnitude of the current through each of the lightbulbs? Explain the pattern.

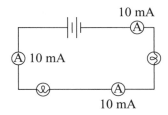

16.3.4 Evaluate the reasoning Your friend says that when two identical lightbulbs are connected in series to each other and then to a battery, the lightbulb connected closest to the negative pole of the battery will be brighter. He explains this by claiming that the second bulb will get fewer electrons because the first bulb will use up some of the electrons. Do you agree or disagree? How can you convince your friend of your opinion? You can use theoretical arguments or perform an experiment to test his suggestion.

16.3.5 Evaluate the reasoning Your friend says that when two identical lightbulbs are connected in series to each other and to the terminals of a battery, the lightbulb closest to the negative pole of the battery will have a greater potential difference across it. She explains it by saying that it will be harder for the electric field to push through to the second bulb after it has already pushed through the first. Do you agree or disagree? How can you convince your friend of your opinion? You can use theoretical arguments or perform an experiment to test her suggestion.

16.3.6 Design an experiment

a. Design an experiment to investigate the relationship between currents through resistors 1, 2, 3, and 4, as shown in the illustration to the right. Describe the experiment and record the results in any format you find appropriate.

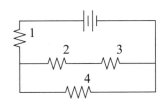

b. Design an experiment to investigate the relationship between the potential difference across resistors 1, 2, 3, and 4 and the potential difference across the battery. Describe the experiment and record the results in any format you find appropriate.

16.3.7 Design an experiment

You have a commercial 9-V battery, a set of resistors, a voltmeter, an ammeter, connecting wires and a switch. Design an experiment to investigate how potential difference across the battery changes as the current through the circuit changes. Make sure you start with the case when the current through the circuit is zero and finish with the maximum possible current (without short circuiting the battery).

a. Draw the circuit for your experiment. Describe the data you plan to collect.

b. Make a table to record the data and after you make the circuit put the data in the table.

c. Describe the pattern you found. How can you explain it? Think of the emf of the battery and its internal resistance.

16.3.8 Derive

Show that the rate at which an electric circuit or an element in the circuit uses electric potential energy (power) is $P = \Delta VI$, where ΔV is the electric potential difference across the circuit or the circuit element and I is the current through that circuit or circuit element. Start with one definition of power P as the rate of electric potential energy ($P = \Delta U_e / \Delta t$) and with the relationship for the electric potential energy change ΔU_q when a charge Δq moves through a potential difference ΔV—that is $\Delta U_q = \Delta q \Delta V$. Combine these two ideas and any others you need to complete the derivation.

16.3.9 Test your idea You have a 9-V battery and two different lightbulbs labeled bulb A and bulb B. You connect them in parallel and see that bulb A is much brighter than bulb B.

a. Use the ideas that you developed in Activity 16.3.8 to explain this observation and then predict what you will observe if you connect the bulbs in series to the same battery.

b. Perform the experiment and record the outcome. Did it match your prediction? If not, revise your explanation to account for the outcome.

16.3.10 Test your ideas Build a circuit consisting of a battery (rated 9 V), a lightbulb, and a switch connected in series. Keep the switch open.

a. Draw the circuit diagram of your circuit below:

b. Predict the potential difference across the battery, across the lightbulb, across a connecting wire and across the switch. Now use a voltmeter to check your predictions. Write down the readings. Discuss any surprising results you found and reconcile them with your prediction.

c. Now close the switch and repeat the experiment. Write down the readings. Do they make sense?

d. Discuss whether Ohm's law in the form of $I = \dfrac{\Delta V}{R}$ applies to a battery and to a switch in an open circuit. Discuss whether Ohm's law applies to a battery, a switch and a connecting wire in a closed circuit.

16.4 | Quantitative Reasoning

16.4.1 Represent and reason Imagine that you have a 9-V battery connected by wires to a lightbulb. Fill in the table that follows and list the assumptions you make in the space below the table. Consider that the negative terminal of the battery is at zero potential.

Draw the circuit.	Draw a qualitative electric potential-versus-position graph.
	V (V) Negative battery terminal Positive battery terminal One side of the bulb Other side of the bulb Negative battery terminal

16.4.2 Represent and reason Complete the table that follows for the circuit in the illustration. List the assumptions you make in the space below the table.

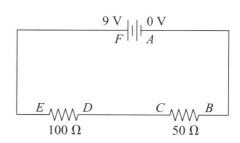

Find the current in the circuit and then calculate the electric potential at each lettered position in the circuit.	Plot the electric potential-versus-position for the circuit.
	V (V) 9 5 0 A B C D E F A

           ©2014 Pearson Education.

16.4.3 Represent and reason The application of Kirchhoff's loop rule for a circuit is shown in the equation that follows.

$$-1.0\,\text{V} - I(2\,\Omega) + 4.0\,\text{V} - I(6\,\Omega) - I(4\,\Omega) = 0$$

Draw a circuit that is consistent with the equation and label the resistors and batteries in the circuit. Draw an arrow and label the electric current.

16.4.4 Represent and reason Complete the last three cells of the table that follows for the circuit shown in the first cell. Do not solve for the currents.

The circuit	Identify currents with arrows and symbols.	Apply the junction rule for junctions A and B.	Apply the loop rule for three different loops.
100 Ω 9 V 1.5 V A ——— B 20 Ω	100 Ω 9 V 1.5 V A ——— B 20 Ω	Junction A: Junction B:	

16.4.5 Represent and reason For the circuit in Activity 16.4.4, determine the current through each branch. *Note:* A branch of a circuit is one part of the circuit where the same electric current flows through each element in that branch.

16.4.6 Represent and reason Complete the last three cells of the table that follows for the circuit shown in the first cell. Do not solve for the currents.

The circuit	Identify currents with arrows and symbols.	Apply the junction rule for junctions A and B.	Apply the loop rule for three different loops.
$\varepsilon_2 > \varepsilon_1$ R_1 A ——— B R_2	$\varepsilon_2 > \varepsilon_1$ R_1 A ——— B R_2	Junction A: Junction B:	

16.4.7 Represent and reason Complete the last three cells of the table that follows for the circuit shown in the first cell. Do not solve for the currents.

The circuit	Identify currents with arrows and symbols.	Apply the junction rule for junctions *A* and *B*.	Apply the loop rule for three different loops.
9 V, 1 Ω / 4.5 V, 0.5 Ω / 7 Ω	9 V, 1 Ω / 4.5 V, 0.5 Ω / 7 Ω	Junction *A*: Junction *B*:	

16.4.8 Reason For the circuit in Activity 16.4.7, determine the current through and the potential difference across each resistor.

16.4.9 Evaluate the solution

The problem: Apply Kirchhoff's loop rule for three loops and the junction rule for one junction for the circuit shown at the right. Do not solve for the electric currents.

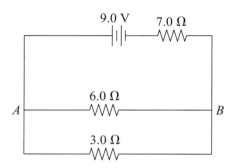

Proposed solution:

$$+9.0 \text{ V} - (7.0 \ \Omega)I - (6.0 \ \Omega)I = 0$$

$$+9.0 \text{ V} - (7.0 \ \Omega)I - (3.0 \ \Omega)I = 0$$

$$- (6.0 \ \Omega)I + (3.0 \ \Omega)I = 0$$

a. Identify any errors in the proposed solution.	b. Provide a corrected solution if there are errors.

16.4.10 Electric circuit Jeopardy You have a circuit consisting of a variety of elements including a 9-V battery (measured as 9 V when you put a voltmeter across it without an external circuit), a switch, and several resistors. You measure current through different circuit elements and the potential difference across them (the same element has the matching voltmeter and ammeter numbers). The results are in the table below. Draw a picture of the circuit where these measurements could have been taken, determine the values of resistances if possible, and show where the voltmeters and ammeters could be located.

Element	Ammeter reading	Voltmeter reading
1	0.071 A	8.86 V
2	0.071 A	7.10 V
3	0.071A	0 V
4	0.035 A	1.76 V
5	0.035 A	1.76 V

16.4.11 Electric circuit Jeopardy You have a circuit with the same 9-V battery as in the previous activity, several resistors, and a switch. You measure current through different circuit elements and the potential difference across them (the same element has the matching voltmeter and ammeter numbers). The results are in the table below. Draw a picture of the circuit where these measurements could have been taken, determine the values of resistances if possible, and show where the voltmeters and ammeters could be located.

Element	Ammeter reading	Voltmeter reading
1	0	9.0 V
2	0	9.0 V
3	0	0

16.4.12 Regular problem What is the potential difference between points A and B if the emfs of the batteries are $\varepsilon_1 = 4.0$ V and $\varepsilon_2 = 1.0$ V and the resistances of the resistors are $R_1 = 10.0 \ \Omega$ and $R_2 = 5.0 \ \Omega$?

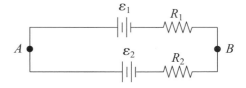

16.4.13 Design an experiment You have a spiral-shaped immersion water heater. Design two experiments to determine the power of the device. Describe the data that you will collect and the mathematical procedure you will use. Examine assumptions in your mathematical procedure. Will they lead to the power estimate higher or lower than actual? If you have the device, conduct the experiments and reconcile the difference between the ratings you obtained.

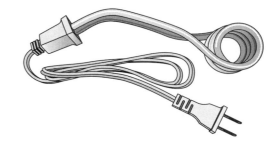

17 Magnetic Forces and Magnetic Fields

17.1 | Qualitative Concept Building and Testing

17.1.1 Observe and find a pattern You have a bar magnet, a horseshoe magnet, and a compass.

a. Perform the experiments and complete the table that follows.

Experiment	Draw the relative orientations of the magnet with marked poles and the compass with marked poles.
Hold a bar magnet horizontally and slowly bring it closer to the compass.	
Repeat the first experiment but reverse the direction of the bar magnet.	
Hold the horseshoe magnet vertically and slowly bring the compass between the poles of the magnet.	

b. Describe in words a pattern between the orientation of the magnet's poles and the orientation of the compass.

c. Describe the same pattern representing a compass with an arrow S → N, as illustrated.

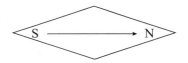

17.1.2 Observe and find a pattern Several experiments are described below.

a. Fill in the table that follows to indicate how magnetic poles are different from or the same as a positively or negatively charged object.

Observations with a magnet	Compare and contrast magnet poles with positively or negatively electrically charged objects.
The north pole of a bar magnet always attracts the south pole of another bar magnet and repels the other bar's north pole.	
Neither the north pole nor the south pole of a magnet exerts a force on a small aluminum ball hanging at the end of a thin, nonconducting thread.	
A foam tube is charged by rubbing so that one end is positive and the other is negative. You find that *both* ends attract the north pole of a bar magnet (and the south pole).	

b. Explain the results of the last experiment in the previous table. Note that a magnet is made of iron—an electric conductor. Start by drawing a picture of the charge distribution on the metal magnet when the positive end of the tube is near the magnet's N pole and again when the tube's negative end is near the magnet's N pole.

17.1.3 Observe and explain Connect a battery, a switch, some wires, and a lightbulb, as shown at the right. The bulb is an indicator of electric current. Make sure that one of the wires in the circuit is aligned along the geographical north–south direction. Place a compass under a north–south-oriented wire with the switch open; there is no current in the circuit. The needle points north.

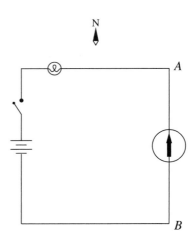

a. Perform the experiments described in the table that follows and record your observations.

Direction of the current	Draw an arrow representing the orientation of the compass when placed below the wire *AB*.	Draw an arrow representing the orientation of the compass when placed above the wire.
Current flows in the north–south ↓ direction in the wire, as shown in the figure.		
If you reverse the battery poles, the current now flows in a south–north ↑ direction in the wire above or below the compass.		

b. Use the thumb of your right hand to represent the direction of the current and your four fingers of the same hand to represent the direction of the compass. Does the orientation of your thumb and fingers describe a pattern between the direction of the current and the orientation of the compass for all of the previous experiments?

c. Come up with a reason why electric current (moving electrically charged particles) might affect the behavior of magnets differently than stationary charged objects would.

17.1.4 Design an experiment Your friend claims that a magnet simply consists of a positive electric charge and a negative electric charge locked at the ends of a metal bar. Design an experiment to test this claim. Predict the outcome.

17.1.5 Observe and explain Imagine that a wire passes up through the page you are reading. Iron filings are sprinkled on the page. We can think of the iron filings as small compasses. The top picture shows the filings when there is no current in the wire. The bottom picture is the arrangement of the filings when there is a significant current in the wire.

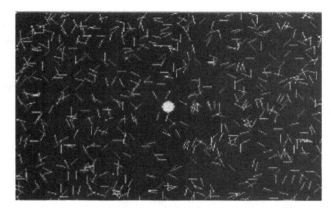

a. Is the picture at the bottom consistent with the results of the experiment in Activity 17.1.3? Explain your answer.

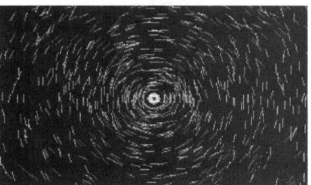

b. Draw a sketch that you think represents the orientation of magnetic field vectors produced by the electric current in the wire at five different points. (*Hint:* Choose the direction of the current as coming out of the page.)

17.1.6 Observe and find a pattern A cathode-ray tube (CRT) is part of a traditional television set or of an oscilloscope. Electrons "evaporate" from a hot filament called the cathode. They accelerate across a potential difference and then move at high speed toward a scintillating screen. The electrons form a bright spot on the screen at the point at which they hit it. A magnet held near the CRT sometimes causes the electron beam to deflect.

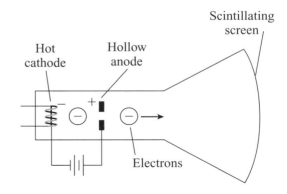

a. Devise a rule for the direction of the force $\vec{F}_{\text{B on E}}$ that the magnet exerts on the moving electrons relative to the direction of their velocity $\vec{v}$ and the direction of the magnetic field $\vec{B}$ produced by the magnet. Use the information provided in the table.

Experiment	Observation
Point the north pole of a magnet at the front of the scintillating screen—opposite the direction the electrons are moving.	Nothing happens to the beam.
Point the north pole of the magnet from the right side (as you face the coming beam) perpendicular to the direction the electrons are moving.	The beam deflects up.
Point the south pole of the magnet from the right side perpendicular to the direction the electrons are moving.	The beam deflects down.
Point the north pole of the magnet down from the top of the CRT, perpendicular to the direction the electrons are moving.	The beam deflects left.

b. Your friend says that the beam of electrons is deflected by the magnet because the electrons are charged particles and the magnet is made of iron. Because all conductors attract electrically charged particles, the experiment above is not related to magnetism. How can you convince your friend that she is mistaken?

17.1.7 Find a pattern A current-carrying wire is placed between the poles of an electromagnet. The direction of the B field lines produced by the magnet ($\vec{B}_{\text{external}}$) is shown in the figure. Invent a rule that relates the directions of the magnetic force $\vec{F}_{\text{B on wire}}$, the directions of the $\vec{B}_{\text{field}}$, and the directions of the current I in the wire.

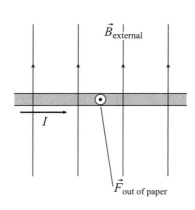

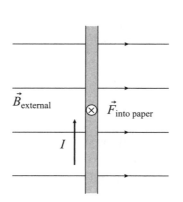

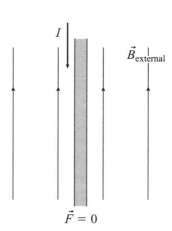

©2014 Pearson Education.

Come up with a rule that relates the directions of $\vec{F}_{B \text{ on wire}}$, $\vec{B}_{\text{external}}$, and I.

17.1.8 Test your idea You have a long wire, a power supply, and a horseshoe magnet. Design an experiment whose outcome you can predict using the rules you came up with in Activities 17.1.6 and 17.1.7. Draw a picture of the apparatus, write your prediction, perform the experiment, and record the outcome.

17.1.9 Represent and reason A rigid wire in the shape of a rectangular loop is shown in the illustration to the right. When the switch in the circuit is closed, there is current around the loop in a clockwise direction. The loop resides in a uniform external magnetic field. Decide the direction of the force exerted on the wires of the loop shown in the figures below and draw it in the figures.

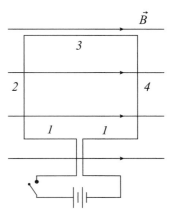

a. $\vec{F}_B$ on side *1*

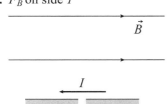

b. $\vec{F}_B$ on side *2*

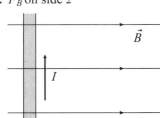

c. $\vec{F}_B$ on side *3*

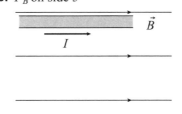

d. $\vec{F}_B$ on side *4*

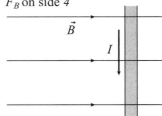

e. Does the magnetic field exert a net nonzero magnetic force on the loop? If so, what is the direction of the net force?

f. Does the field exert a net nonzero magnetic torque on the loop? If so, how does this torque tend to cause the orientation of the loop to change (assuming it is initially at rest)?

17.1.10 Represent and reason A rigid wire in the shape of a rectangular loop is shown at the right. When the switch in the circuit is closed, current flows up side *2*, across side *3*, and down side *4*. The loop's surface is perpendicular to the page and resides in an external magnetic field. The field's direction is parallel to the page and perpendicular to the surface of the loop. Decide the direction of the force exerted on the wires of the loop shown in the figures below and draw it in the figures.

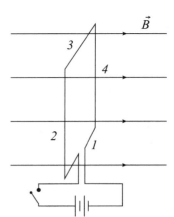

a. $\vec{F}_B$ on side *1*

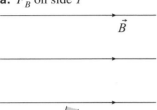

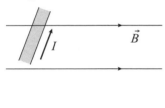

b. $\vec{F}_B$ on side *2*

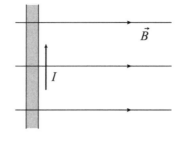

c. $\vec{F}_B$ on side *3*

d. $\vec{F}_B$ on side *4*

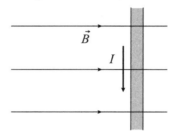

e. Does the magnetic field exert a net nonzero magnetic force on the loop? If so, what is the direction of the net force?

f. Does the field exert a net nonzero magnetic torque on the loop? If so, how does this torque tend to cause the orientation of the loop to change (assuming the loop is initially at rest)?

17.2 | Conceptual Reasoning

17.2.1 Represent and reason For each situation represented in the illustration, decide if the external magnetic field (source) exerts a nonzero magnetic force on the short section of the current-carrying wire (test object). If the magnetic force is not zero, indicate the direction of the magnetic force by drawing it in the figures that follow. (*Note:* There must be other sections of wire, not shown, that are connected to these wires.)

a.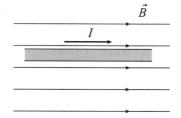

b.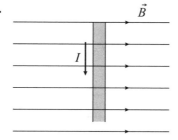

c.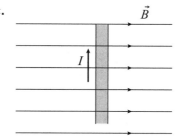

d.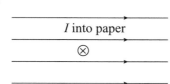

e.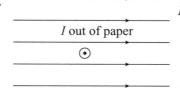

f.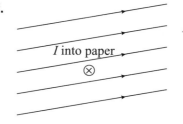

g.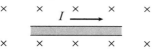

h.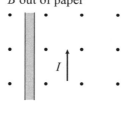

17.2.2 Represent and reason
For each situation depicted in the table that follows, find the direction of the unknown physical quantity. Draw in the directions in the figures.

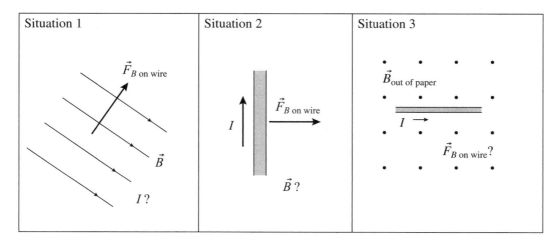

17.2.3 Represent and reason For each situation below, decide if a nonzero magnetic force is exerted on the moving electric charge (test object). If the force is not zero, draw in the direction of the magnetic force on the figures that follow.

a.
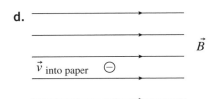

b.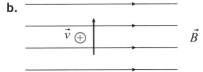
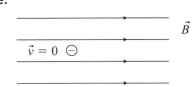

c.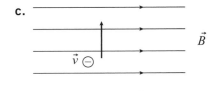
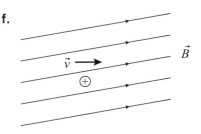

d.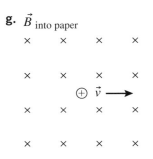

e.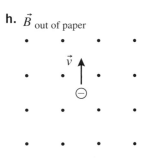

f.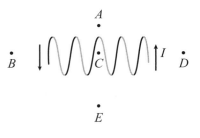

g. $\vec{B}$ into paper
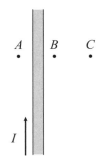

h. $\vec{B}$ out of paper
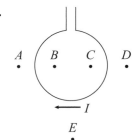

17.2.4 Represent and reason There is current in each of the wires shown in the illustration. Determine the direction of the magnetic field created by the current at the points indicated and draw it with an arrow, a dot (out of the page), or a cross (into the page). The objects indicated in the illustration are sources of a magnetic field.

a.

A B C
• • •

I

b.

A B C D
• • • •

I

E
•

c.

A
•

B   C   I   D
•   •   •

E
•

17.2.5 Pose a problem Pose a qualitative problem based on the situation shown in the illustration. The airplane is flying through Earth's magnetic field.

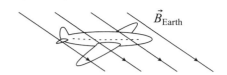

17.2.6 Represent and reason Two wires are parallel to each other. Wire 1 has electric current going into the page, and wire 2 has electric current coming out of the page. Fill in the table that follows to emphasize the importance of choosing a source of a field and a system on which the field exerts a force.

Draw in B field lines $\vec{B}_1$ produced by the current I_1 in wire 1 (the source of the field). Be sure to include a line passing through wire 2.	Noting the field $\vec{B}_1$ passing through wire 2, draw the direction of the magnetic force $\vec{F}_{1 \text{ on } 2}$ that wire 1's magnetic field $\vec{B}_1$ exerts on wire 2 (the system).	Draw in B field lines $\vec{B}_2$ produced by the current I_2 in wire 2 (the source of the field). Be sure to include a line passing through wire 1.	Noting the field $\vec{B}_2$ passing through wire 1, draw the direction of the magnetic force $\vec{F}_{2 \text{ on } 1}$ that wire 2's magnetic field $\vec{B}_2$ exerts on wire 1 (the system).
1 ⊗ 2 ⊙ I_1 in	1 ⊗ 2 ⊙ I_2 out	1 ⊗ 2 ⊙ I_2 out	1 ⊗ 2 ⊙ I_1 in

17.2.7 Represent and reason There is an electric current through a horizontal bar that hangs from two thin side wires (see the side view at the right). In what direction should an external magnetic field point so that the magnetic force that the magnetic field exerts on the bar helps support the bar? Explain. Draw a force diagram.

Side view

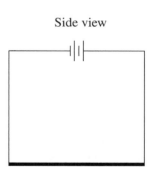

17.2.8 Represent and reason A word description and sketch of a hypothetical model of an atom is shown below.

a. Fill in the table that follows. Construct a physical representation that treats the atom like a circular loop of wire with an electric current in a magnetic field. *Hint:* The electron moves about the nucleus very fast, in effect producing a circular electric current about the nucleus.

Word description	Sketch	Construct a physical representation.
In the Bohr model of the hydrogen atom, a negatively charged electron moves in a circular orbit around a positively charged proton nucleus. Suppose one of these hydrogen atoms is in a magnetic field.	$\vec{B}$ Electron ⊖ ⊕	

b. Does the field produce a torque on the atom? If so, indicate the axis of rotation and the direction the atom would tend to turn about that axis.

17.2.9 Represent and reason Does the magnetic field exert a nonzero torque on the current loop in each case pictured below? If so, and if the loop is initially at rest, which way would the magnetic torque cause the loop to start turning? For each case draw in the forces on two opposite sides of the loop and show the direction of the net torque. (Current loops in a., b., and d. are perpendicular to the page.)

a.

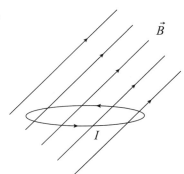

b. $\vec{B}$ into page

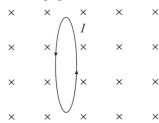

c. $\vec{B}$ into page

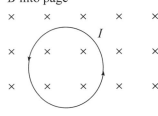

d.

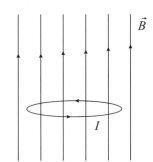

e.

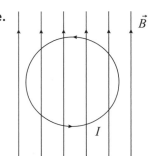

17.2.10 Reason A positively charged proton is ejected into space during a supernova, an explosion that occurs during the gravitational collapse of a large star near the end of its life. The proton travels through space for millions of years and finally reaches the magnetic field that surrounds Earth; we call the proton a cosmic ray. Sketch the path of the proton as it travels in Earth's magnetic field. Show its path as it might be seen from below Earth (see the person in the drawing).

Side view

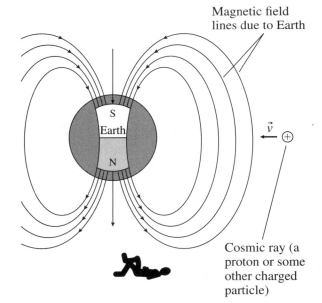

Magnetic field lines due to Earth

Cosmic ray (a proton or some other charged particle)

17.3 Quantitative Concept Building and Testing

17.3.1 Find a pattern The table below provides data concerning the magnitude of the magnetic force $\vec{F}_{\text{B on w}}$ exerted on a segment of a current-carrying wire by an external magnetic field as the following quantities are changed: (1) the magnitude of the external magnetic field $\vec{B}$, (2) the magnitude of electric current I, (3) the length of the segment of the current-carrying wire L, and (4) the direction of the electric current relative to the direction of the magnetic field.

Magnitude of the magnetic field $\vec{B}$ (T)	Current I in the wire (A)	Length L of the wire (m)	Angle θ between current direction and $\vec{B}$ field	Magnitude of the magnetic force F_B exerted on the wire (N)
1B	I	L	90°	F
2B	I	L	90°	2F
3B	I	L	90°	3F
B	1I	L	90°	F
B	2I	L	90°	2F
B	3I	L	90°	3F
B	I	1L	90°	F
B	I	2L	90°	2F
B	I	3L	90°	3F
B	I	L	0°	0
B	I	L	30°	0.5F
B	I	L	90°	F

Devise a rule relating the magnitude of the magnetic force F_B to these quantities.

17.3.2 Find a pattern The table below provides data concerning the magnitude of the magnetic force exerted on a moving charged particle by a magnetic field as the following quantities are changed: (1) the particle's speed, (2) the magnitude of the magnetic field, and (3) the direction of the particle velocity relative to the magnetic field.

Magnitude of the magnetic field $\vec{B}$ (T)	Charge of the moving particle	Speed v of the moving particle (m/s)	Angle θ between the velocity $\vec{v}$ and the $\vec{B}$ field	Magnitude of the magnetic force F_B exerted on the particle (N)
1B	q	v	90°	F
2B	q	v	90°	2F
3B	q	v	90°	3F
B	q	v	90°	F
B	2q	v	90°	2F

(continued)

Magnitude of the magnetic field $\vec{B}$ (T)	Charge of the moving particle	Speed v of the moving particle (m/s)	Angle θ between the velocity $\vec{v}$ and the $\vec{B}$ field	Magnitude of the magnetic force F_B exerted on the particle (N)
B	$3q$	v	90°	$3F$
B	q	v	90°	F
B	q	$2v$	90°	$2F$
B	q	$3v$	90°	$3F$
B	q	v	0°	0
B	q	v	30°	$0.5F$
B	q	v	90°	F

Devise a rule relating the magnitude of the force to these quantities.

17.3.3 Predict and test Assemble the apparatus shown in the illustration (a thick horizontal wire hanging from support wires on each side) and place a horseshoe magnet on a scale. Hang the apparatus between the poles of the magnet. Observe what happens to the reading of the scale when you turn on the current. Use the rule that you devised in Activity 17.3.2 to predict what will happen to the reading of the scale when you double the magnitude of the current. Fill in the table that follows.

Side view

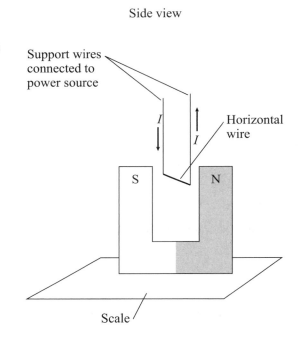

Support wires connected to power source

Horizontal wire

S   N

Scale

Record the reading of the scale with the magnet on it (no current in the wire).	Explain in words why the reading of the scale changes when the current is turned on.	Write a procedure to predict the reading of the scale when the current is doubled. List your assumptions.	Perform the experiment, record the outcome, and reconcile the outcome with the prediction.
		Procedure: Assumptions:	

17.3.4 Reason A galvanometer is a device that serves as a basis for an ammeter and a voltmeter. The galvanometer consists of a coil hanging between the poles of a horseshoe magnet. The coil is supported by a rod that can turn in a balljoint. A spring opposes its turning. A needle, attached to the rod, changes its orientation as the rod turns. The greater the current flowing through the coil, the greater the torque exerted on it by the magnetic field of the magnet, and the more the needle deflects. Discuss how one can make an ammeter and a voltmeter out of the same galvanometer. Imagine that you have resistors of different resistances.

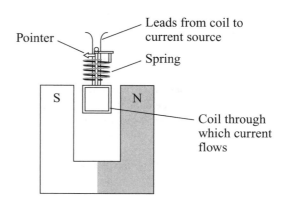

17.4 | Quantitative Reasoning

17.4.1 Represent and reason The mass-detecting part of a mass spectrometer is described below in multiple ways. Starting with basic equations, devise a mathematical expression for the particle mass.

Words	Sketch	Physical representation	Mathematical representation
An ion with mass m and charge $+e$ leaves a velocity selector moving at speed v. It then moves in a half circle in a magnetic field that is perpendicular to the plane of its motion. At the end of this trip, it is detected. The radius of the circle can be used to determine the mass of the ion.	Top view Velocity selector $\vec{B}$ out of paper $2r$ $\oplus$ $\vec{v}$ Detector	Radial ←— $\vec{F}_m$ ●	

17.4.2 Represent and reason The speed of blood flow in an artery can be measured using our knowledge of a magnetic field and an electric field. The process is described below in words. Represent the process in other ways.

Description in words
Positive and negative ions move with blood in an artery through a magnetic field that is perpendicular to the blood's velocity. The magnetic force causes some positive ions to accumulate on one wall of the artery and negative ions on the other wall. This charge separation causes an electric field that opposes further charge separation. Derive an expression for the speed of the blood in terms of the electric and magnetic fields.

Draw a sketch.	Write a mathematical representation.	Draw a physical representation.

17.4.3 Regular problem What happens to a cosmic-ray proton flying into the Earth's atmosphere at a speed of about 10^7 m/s? The magnitude of the Earth's $\vec{B}$ field is approximately 5×10^{-5} T. The mass m of a proton is approximately 10^{-27} kg. Consider three cases: The proton enters the Earth's atmosphere parallel to the $\vec{B}$ field, perpendicular to the field, and at a 30° angle.

17.4.4 Equation Jeopardy Two processes are represented mathematically below, using Newton's second law. Fill in the table that follows to describe the processes in other ways.

Mathematical representation	Construct a physical representation.	Sketch the situation.	Write a description of the problem in words.
$(1.6 \times 10^{-19} \text{ C})$ $(2.0 \times 10^7 \text{m/s}) B$ $= (1.67 \times 10^{-27} \text{ kg})$ $(2.0 \times 10^7 \text{ m/s})^2/(6000 \text{ m})$			
$0.020 \text{ N} = (0.020 \text{ A})$ $(0.10 \text{ T}) (20 \text{ m}) (0.50)$			

17.4.5 Evaluate the solution

The problem: You are playing a video air-hockey game in which a hockey puck of mass m with electric charge $+q$ leaves a velocity selector traveling at speed v on a horizontal frictionless surface. You are to write an expression for the magnitude B of a magnetic field and decide on its direction so that it bends the puck in a curving half circle (toward the bottom of the screen) and it hits a target a distance $2R$ below the place the puck left the velocity selector.

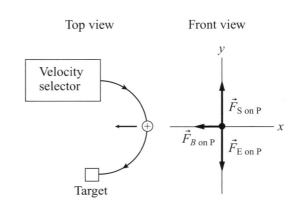

Proposed solution: The situation is pictured at the right, along with a force diagram for the puck as seen in the plane of its motion at the instant the puck reached the halfway point around the half circle. The acceleration direction is indicated.

Mathematical representation:

$$\Sigma F_{rad} = mv^2/R \quad \text{or} \quad qvB \sin 90° = mv^2/R$$

Consequently, $B = mv^2/qR$ and the field points to the left.

a. Identify any missing elements or errors in the solution.

b. Provide a corrected solution if there are errors.

17.4.6 Evaluate the solution

The problem: You wish to impress your friends with your mystical powers and decide to build a small object that you can cause to levitate at the dinner table. How will you do this?

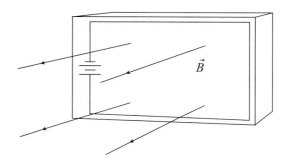

Proposed solution: You decide to place a 0.20-cm-high, 1.0-cm-long rectangular loop of resistance 0.30 Ω inside a 2.0-g box. A tiny 1.5-V battery sends a current through the coil (after a hidden switch is turned on). When you place your hands on each side of the box (with a 2.0-T field caused by the magnets you hide between your fingers), the force of the magnetic field causes the top and bottom parts of the current-carrying wire in the loop to rise up into the air.

Mathematical representation:

$$\Sigma F_y = 2ILB = mg$$

Note that the magnetic force is exerted on the two long horizontal sections of the loop wire. We check to see if the two sides of the equation are equal:

The left side:

$2(1.5 \text{ V}/0.30 \ \Omega)(1.0 \text{ cm})(2.0 \text{ T}) = 20 \text{ N}$

The right side:

$(2.0 \text{ g})(10 \text{ m/s}^2) = 20 \text{ N}$

a. Identify any missing elements or errors in the solution.

b. Provide a corrected solution or missing elements if you find errors.

17.4.7 Evaluate You work in the complaint office of an auto manufacturer. A customer makes a complaint. Fill in the table that follows to describe the process in other ways and decide how you, an employee in the complaint office, will respond.

Description of the argument in words	A customer says that while driving through Earth's magnetic field, a potential difference developed from one end of her car to the other. The car discharged, causing her to lose control and run off the highway into a ditch. Decide the action you should take. Be sure to use your physics knowledge to help complete the recommendation—people have much more confidence in decisions based on physics, don't they?
Sketch the situation.	Construct a physical representation.
Write a mathematical representation.	Evaluate the argument and provide a recommendation.

17.4.8 Evaluate A neighborhood group asks your advice concerning the hazard of a new electric line that is to go into your neighborhood to provide power for streetlights. The neighbors have heard that magnetic fields caused by power lines may cause cancer to people living near the lines. Use your knowledge of physics to try to help you make a decision about whether the proposed electric line in your neighborhood will cause health hazards. Indicate any assumptions you made.

17.4.9 Reason Suppose that the rail for a train had an electric current traveling through it and that the train had a large coil such as that shown on the right. Can the magnetic field produced by the rail exert a magnetic force on the train that helps support it—like magnetic levitation? Explain. *Hint:* The magnetic field strength decreases with distance from the current-carrying wire.

18 Electromagnetic Induction

| **Qualitative Concept Building and Testing**

18.1.1 Observe and find a pattern The table that follows describes six experiments involving a galvanometer, a bar magnet, and a coil. Perform the experiments and record the outcomes.

Experiment	Illustration	Outcome
a. Hold the magnet motionless in front of the coil, with any orientation.		
b. Hold the magnet perpendicular to the coil with the N pole facing the coil. Move the magnet quickly toward the coil. Then pull it away quickly.		

(continued)

Experiment	Illustration	Outcome
c. Repeat experiment 2, but this time with the S pole facing the coil.		
d. Align the magnet in the same plane as the coil, and move either pole toward or away from the coil.		
e. Hold the magnet in front of the coil and rotate it 90° as shown. (The magnet starts out perpendicular to the coil and ends up parallel to it.)		
f. Position the magnet as in experiment 2, but this time grasp the sides of the coil and collapse the coil quickly. Then pull it back open.		

Devise a rule that summarizes when a current is induced in a coil.

18.1.2 Test your idea Four experiments using a galvanometer, a switch, and a coil are described in the table that follows. Use the rule or rules you devised in Activity 18.1.1 to predict if there should be an induced current in the coil that is not connected to a battery shown in the following illustrations, as detected by the galvanometers.

a. Fill in the table that follows.

Experiment	Illustration	Predict the outcome.	Perform the experiment and record the outcome.
The current in the left coil increases just after the switch is closed.	Switch Galvanometer		
The current in the left coil increases just after the switch is closed. The coils are perpendicular.	Switch Galvanometer		
The current in the left coil decreases just after the switch is opened.	Switch Galvanometer		
The switch in the left coil is closed (steady current) as the right coil moves toward and above the left coil.	Galvanometer		

b. If necessary, revise the rule you developed in Activity 18.1.1.

18.1.3 Test your idea The switch in the left coil pictured in the illustration at right is closed (there is a steady current in the left coil), and both coils move right at the same velocity.

Galvanometer

a. Predict whether the galvanometer will register an induced current as the two coils are moving. Explain your prediction.

b. Perform the experiment and record the outcome. If the outcome contradicts your prediction in part a, discuss whether you consistently used the rule developed in earlier activities to make the prediction.

18.1.4 Test an idea David says that the size of the magnet determines whether a current can be induced in a coil. You want to convince him that his idea is not correct. One way to do it is to design an experiment whose outcome might contradict a prediction based on David's idea.

a. Fill in the table that follows.

Describe an experiment to test David's idea.	Make a prediction of its outcome based on David's idea.	Perform the experiment and record the outcome.

b. Discuss whether David will be convinced by your results.

18.1.5 Observe and find a pattern The table that follows describes five new experiments using a galvanometer, bar magnets, and a coil. Perform the experiments and compare the outcomes to the described outcomes. The outcomes of the experiments are included.

Experiment	Illustration	Outcome
a. Position a magnet perpendicular to the coil and move it slowly toward the coil. Repeat the experiment, moving the magnet quickly.		The quicker the magnet's motion, the stronger the induced current.
b. Position a smaller magnet perpendicular to the coil and move it slowly toward the coil. Repeat the experiment using a bigger magnet.		The bigger magnet induces a stronger current than the smaller magnet when they move at the same speed with respect to the coil.
c. Move a magnet perpendicular to the coil. Then move it so that it makes an angle with the plane of the coil. Keep the speed the same.		When the magnet moves perpendicular to the coil, the strongest current is induced.

d. Make a small coil and a large coil. Move the magnet toward each.	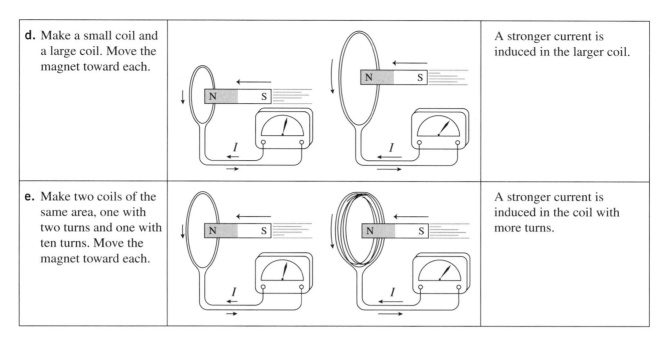	A stronger current is induced in the larger coil.
e. Make two coils of the same area, one with two turns and one with ten turns. Move the magnet toward each.		A stronger current is induced in the coil with more turns.

Devise a rule that relates the *magnitude* of the induced current to various properties of the magnet, its motion, and properties of the coil.

18.1.6 Observe and find a pattern The table repeats three earlier experiments that used a galvanometer, a bar magnet, and a coil and in which a current was induced. The direction of the induced current is shown in the illustrations.

a. Fill in the table that follows.

Experiment	Draw $\vec{B}_{ext}$ field vectors, and $\Delta\vec{B}_{ext}$ vectors through the coil caused by the moving magnet. Indicate whether the external magnetic flux through the coil is decreasing or increasing. Draw $\vec{B}_{ind}$ field vectors due to the induced current.

(*continued*)

Experiment	Draw $\vec{B}_{ext}$ field vectors, and $\Delta\vec{B}_{ext}$ vectors through the coil caused by the moving magnet. Indicate whether the external magnetic flux through the coil is decreasing or increasing. Draw $\vec{B}_{ind}$ field vectors due to the induced current.

b. Use the data in the table to devise a rule relating the direction of the induced current in the coil and the change of external magnetic flux through it. *Hint:* (1) Draw the $\Delta\vec{B}_{ext}$ field vectors of the bar magnet and make a note of whether the flux due to this magnet is increasing or decreasing through the coil. (2) Then draw $\vec{B}_{ind}$ vectors as a result of the induced electric current. (3) Compare the direction of $\vec{B}_{ind}$ vectors to the $\Delta\vec{B}_{ext}$ field vectors of the bar magnet when the flux through the coil increases and (4) when the flux decreases.

c. How does the direction of the induced current in a coil relate to the change of external magnetic flux through it?

18.2 | Conceptual Reasoning

18.2.1 Reason For each situation shown in the table that follows, use the rules devised and tested in Section 18.1 to predict if a current is induced through the resistor attached to the loop. If a current is induced, indicate the direction of that induced current.

Experiment	Predict if a current is induced; explain your prediction.	If you predict that a current is induced, what is the direction of the current?
a. The loop is perpendicular to the page.		

b. The loop is perpendicular to the page and the magnet turns 90°.

c. The loop is in the plane of the page.

d. The loop, perpendicular to the page, is pulled upward so that it collapses.

Pull up here.

Remains stationary

Remains stationary

e. The switch in the left circuit is closed, and the current increases abruptly.

f. There is a steady current in the left circuit.

g. The circuit on the left is rotated 90°.

(continued)

Experiment	Predict if a current is induced; explain your prediction.	If you predict that a current is induced, what is the direction of the current?
f. The switch in the left circuit is opened, and the current decreases abruptly. A ⌁WWW⌁ B		

18.2.2 Represent and reason The rectangular loop with a resistor is pulled at constant velocity through a uniform external magnetic field that points into the paper in the regions shown in the illustration with the crosses (×).

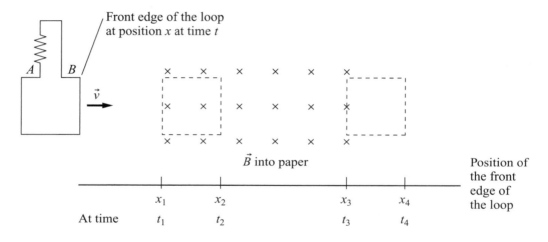

Complete the table that follows to determine qualitatively the shape of the induced current-versus-time graph.

a. Draw a qualitative flux-versus-time graph for the process (positive in and negative out).	Flux Φ t_1 t_2 t_3 t_4 — Time t
b. Draw a qualitative induced magnetic field-versus-time graph for the process.	Induced magnetic field B_{in} t_1 t_2 t_3 t_4 — Time t

c. Draw a qualitative induced current-versus-time graph for the process.	Induced current I_{in} ⊢————┼————┼————————┼————┼——— Time t 　　　　　　t_1　　t_2　　　　　t_3　t_4

18.2.3 Evaluate the solution

The problem: The magnetic field through the square coil shown in the illustration is at first steady and large (there are many turns in the coil, but only one is shown). The field then decreases to zero in about 1.0 s. A bulb connected to the ends of the coil indicates an induced current in the circuit. When are you most likely to observe light from the bulb?

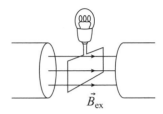

Proposed solution: A steady light will come from the bulb when the field is steady and large. The brightness of the light will decrease as the field decreases. There is no light when the magnetic field becomes zero.

a. Identify any errors in the proposed solution.

b. Provide a corrected solution if there are errors.

18.3 | Quantitative Concept Building and Testing

18.3.1 Observe and explain In the table that follows, the results of four experiments are shown in which a changing magnetic field produced by an electromagnet passes through a loop, as illustrated to the right. This changing $\vec{B}$ field causes a changing flux Φ through the loop and an induced current I_{ind} around the loop of resistance R. The product $I_{ind}R$ is also plotted as a function of time.

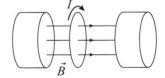

a. Draw a fourth graph in each table column that shows the induced emf ε_{ind} in the loop.

Coil resistance is 1.0 Ω.	Coil resistance is 3.0 Ω.	Coil resistance is 2.0 Ω.	Coil resistance is 6.0 Ω.

b. Devise a relationship between $\Delta\Phi/\Delta t$ and ε_{ind}. Do not forget the sign!

18.3.2 Observe and explain The analysis of the following experiment will help you devise an expression for the potential difference produced in a loop moving into, through, and out of a magnetic field. This is called *motional emf.*

Experiment	Analysis	Analysis
A metal bar of length L moves at constant velocity through a magnetic field that points into the paper (the crosses).	**a.** Indicate on the bar how the magnetic force exerted by the field on charges in the bar redistributes charges in the bar.	**b.** This charge redistribution, which occurs quickly, produces an electric field inside the bar that prevents further charge redistribution. Draw the $\vec{E}$ field lines.

Analysis	Analysis	Analysis
c. Apply Newton's second law for a charge in the middle of the bar—now in an $\vec{E}$ field and a $\vec{B}$ field.	**d.** Use the expression that relates electric field E_y and the potential difference over a distance $\Delta V/L$ with the previous results to determine an expression for the potential difference (emf) across the ends of the bar.	**e.** Below, a rectangular metal conductor is depicted entering the magnetic field, residing completely in that field, and leaving the field described above. Use the results of parts a and b to draw on the left and right vertical parts of the rectangle the charge distributions due to the force exerted by the magnetic field on the electrically charged particles in the metal. *Note:* The magnetic field only exerts forces on charges in the metal parts that are in the field, not on parts outside the field. Will there be electric current in the conductor? If so, indicate the direction.
f. Use the results of part d to write an expression for the potential difference induced around the rectangular metal conductor while entering the field.	**g.** Use the results of part d to write an expression for the potential difference induced around the rectangular metal conductor when completely in the field. Note that there are now charge distributions that cancel each other.	**h.** Use the results of part d to write an expression for the potential difference induced around the rectangular metal conductor while leaving the field.

18.3.3 Observe and explain Repeat the previous activity, only this time use the ideas of flux and induced emf. When you are finished, check to see if the results are consistent with those in Activity 18.3.2.

Experiment	Analysis
The same rectangular metal conductor as in Activity 18.3.2e is entering the magnetic field, moving completely in the magnetic field, and leaving a magnetic field that points into the paper.	**a.** Draw a graph showing the magnetic flux through the opening of the metal conductor as a function of time while the rectangle is entering the magnetic field, moving completely in the magnetic field, and leaving the magnetic field. Then use the flux graph to make a graph of the induced emf for the same time interval.

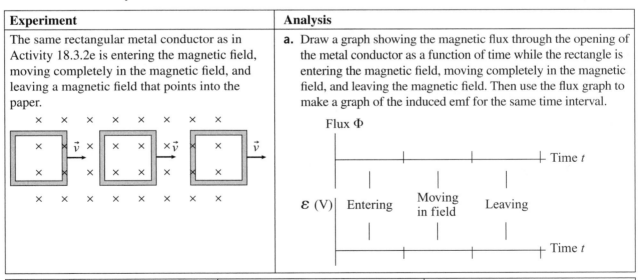

Analysis	Analysis	Analysis
b. Write an expression for the changing flux as the rectangle enters the field. Then use this expression to determine the emf while the rectangle is entering the field.	**c.** Write an expression for the changing flux as the rectangle is completely in the field. Then use this expression to determine the emf while the rectangle is in the field.	**d.** Compare the expressions in parts b and c with the expressions determined in 18.3.2f and g. (We are skipping the calculation for when the rectangle is leaving the field—it's a little more messy.)

18.4 | Quantitative Reasoning

18.4.1 Reason The magnitude of the magnetic field in each situation described below is 0.50 T. For each situation in this table, write an expression for the magnetic flux through the loop of radius r.

Situation	Write an expression for the flux.	Situation	Write an expression for the flux.
Loop and $\vec{B}$ in the plane of the paper 		Loop perpendicular to the paper and $\vec{B}$ in the plane of the paper 	
Loop in the plane of the paper $\vec{B}$ out of paper 		Square loop of side A in the plane of the paper. $\vec{B}$ into the paper 	

18.4.2 Represent and reason Four situations are shown in which the external flux through a loop is plotted as a function of time. In the table that follows, draw another graph that shows the induced emf in the loop as a function of time.

Situation 1	Situation 2

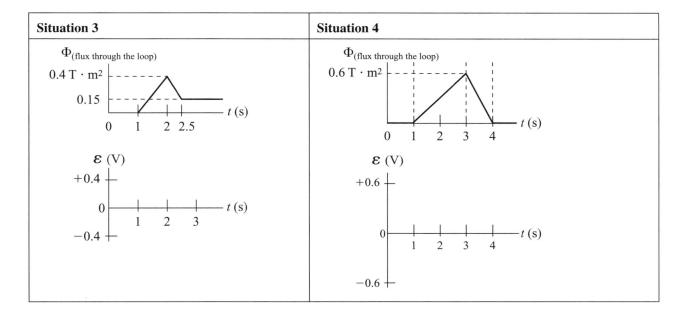

Situation 3	Situation 4

18.4.3 Equation Jeopardy Write a problem in words and construct a sketch for a phenomenon involving electromagnetic induction that is described by each equation below (there is more than one possibility). Provide all the details for this phenomenon.

Mathematical description	Write a description in words of a problem that is consistent with the equation.	Sketch a situation the problem might describe.
$\varepsilon = -20[\pi(0.10 \text{ m})^2] \cos 37° \times (0 - 0.40 \text{ T})/(2.0 \text{ s} - 0)$		
$\varepsilon = -4(0.40 \text{ T}) \cos 0° \times [0 - (0.10 \text{ m} \times 0.20 \text{ m})]/(2.0 \text{ s} - 0)$		

18.4.4 Evaluate the solution

The problem: A single 0.10-m × 0.10-m square loop is between the poles of a large electromagnet. The surface of the loop makes a 53° angle with respect to the magnetic field. The magnetic field varies as shown in the illustration. Determine the induced emf produced by the loop.

Proposed solution:

Sketch and translate

See the drawing at the right.

Simplify and diagram

The magnetic flux Φ through the loop increases as the magnetic field increases (see the graph). The emf ε has the same shape, only with a negative sign (the lower graph line).

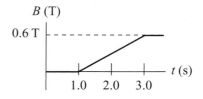

Represent mathematically and solve

$$\varepsilon_{\text{ind av}} = N\Delta(BA\cos\theta)/\Delta t = NA\cos 53°\left[(B_f - B_i)/(t_f - t_i)\right]$$
$$= (1)(0.10\text{ m})^2\{0.60[(0.60\text{ T} - 0)/(3.0\text{ s} - 0)]\}$$
$$= 1.2 \times 10^{-3}\text{ T}\cdot\text{m}^2/\text{s} = 1.2 \times 10^{-3}\text{ V}$$

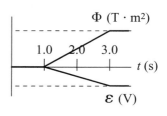

a. Identify any errors in the solution.

b. Provide a corrected solution if there are errors.

18.4.5 Evaluate the solution

The problem: A single square loop is between the poles of a large electromagnet with its surface perpendicular to the external magnetic field. The magnetic field decreases steadily to zero. Determine the direction of the induced current in the loop.

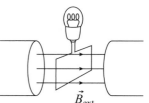

Proposed solution: The induced magnetic field (shown with dashed arrows) must point left to oppose the external magnetic field (the solid arrows). A clockwise induced current (the curved arrow) will produce this induced magnetic field.

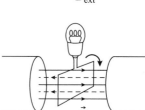

a. Identify any errors or missing elements in the proposed solution.

b. Provide a corrected solution if there are errors.

18.4.6 Pose a problem
Design a problem that involves a graphical representation of magnetic flux-versus-time. You can provide the graph and ask for some other information based on the graph or provide other information and construct the graph as part of the problem solution.

18.4.7 Pose a problem
Design a problem that involves a graphical representation of induced emf-versus-time. You can provide the graph and ask for other information based on the graph or provide other information and construct the graph as part of the problem solution.

18.4.8 Regular problem
A horizontal bar is pulled at a constant velocity through a downward-pointing magnetic field. The bar slides on two horizontal, frictionless metal rails moving away from a resistor connected between their ends. Derive an expression for the induced current through

the resistor of resistance R in terms of any or all quantities that you choose to include in a sketch of this system. Be sure to identify the quantities in the sketch.

a. Draw top-view sketches showing the rails and the bar location at an initial time and at a later time. Include symbols for quantities involved in the problem.

b. Describe the assumptions you are making.

c. Construct labeled graphs for the process below.

Plot the flux through the area surrounded by the rails, moving bar, and the resistor at the end of the rails as a function of time.

Consistent emf-versus-time graph

d. Represent flux and emf mathematically.

e. Combine the mathematical representation with Ohm's law to get the desired expression for the current.

18.4.9 Regular problem Loop *KLMN* is made of metal rods, where rod *KL* slides at a constant speed on the side rods *NK* and *ML* in the direction indicated by the arrow. A constant external magnetic field either points down into the loop (the crosses) or up out of the loop (the dots). For each situation, use two different methods to determine the direction (not the magnitude) of the electric current in loop *KLMN*.

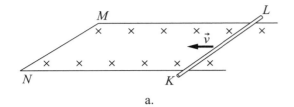

a.

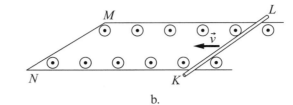

b.

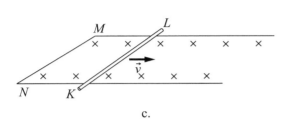

c.

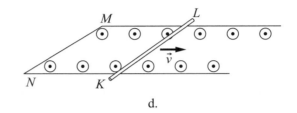

d.

a. Use the magnetic force on free charges in the rod to find charge separation on the rod, the electric force that this charge separation produces that prevents additional charge separation, and the relationship between the electric field caused by the charge separation and electric potential difference to determine which side of the moving rod is at higher electric potential and which way electric current flows around the loop *KLMN*. Draw the direction of the current in the figure above.

b. Next use Lenz's law and right-hand rule for the magnetic field to determine the direction of the induced current. Draw the direction of the current determined with this method in the figure above using a different-colored pencil.

c. Do these two approaches agree? Explain.

18.4.10 Regular problem The Dreamworld's Tower of Terror at the Gold Coast in Australia is the fastest, tallest thrill ride in the Southern Hemisphere. Its L-shaped track has a smooth curve between the horizontal section of the track and the vertical section. The 15-passenger cart goes from 0 to 100 mph in 7 s on the horizontal section of the track and then coasts up to the top of the track, like a ball thrown up into the air. The cart then falls back down and is stopped by eddy-current braking on the horizontal track where it started.

a. Describe how this braking might work.

b. Estimate the distance traveled while the rollercoaster cart gets up to speed.

c. Estimate the acceleration in *g*'s while the rollercoaster cart starts.

d. Estimate the maximum height reached by the cart.

18.4.11 Pose a problem An airplane flies at a speed of 900 km/h. The distance between the tips of its wings is about 12 m. The vertical component of Earth's magnetic field is about 5×10^{-5} T. Pose a problem using this information.

19 Vibrations

19.1 | Qualitative Concept Building and Testing

19.1.1 Observe and find a pattern Try the following simple experiments and describe common patterns concerning the behavior of the block.

a. Fill in the table that follows.

Experiment	Record your observations.
Tie a string to a small, heavy block and let the block hang freely. Now pull the block to the side and release it.	
Hang a heavy block from a spring, pull the block down, and release it.	

b. Identify patterns common to both experiments.

19.1.2 Explain In Activity 19.1.1 you found that both the block on a string and the block on a spring had repeatable motion, either back and forth or up and down. The blocks moved about the place where they resided when not vibrating—that is, about the *equilibrium position.* Explain why each block returns to this equilibrium position, first moving in one direction and then a short time later in the opposite direction. To help your thinking, draw force diagrams for the block when on each side of the equilibrium position.

19.1.3 Test your idea Clamp the top of a spring to a ring stand. Attach a horizontal metal bar to the bottom of the spring. Suppose you rotate the bar about 90° to the side perpendicular to the spring, so that it twists the spring. Predict what happens to the bar when you release it. Be sure to identify the equilibrium position for the bar and the reason it moves as it does.

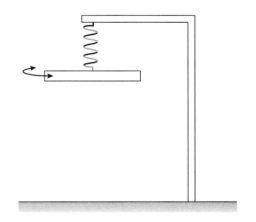

19.2.1 Represent and reason The cart in the figure is attached to a special spring that can stretch and compress equally well. The spring is very light. The cart and spring rest on a low-friction horizontal surface. The cart is pulled to position I and then released. It moves to position V, where it then reverses direction and returns again to position I. It repeats the motion. Represent with motion diagrams and force diagrams the cart's motion between the points indicated in the table that follows.

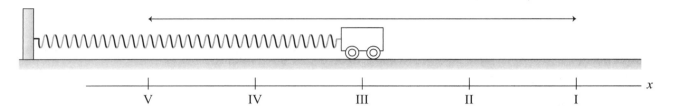

Draw a motion diagram for motion between points I–III.	Draw a motion diagram for motion between points III–V.	Draw a motion diagram for motion between points V–III.	Draw a motion diagram for motion between points III–I.
Draw a force diagram for point I, cart moving left.	Draw a force diagram for point III, cart moving left.	Draw a force diagram for point V, cart moving left.	Draw a force diagram for point II, cart moving left.
Draw a force diagram for point I, cart moving right.	Draw a force diagram for point III, cart moving right.	Draw a force diagram for point V, cart moving right.	Draw a force diagram for point II, cart moving right.

a. Do the force diagrams depend on whether the cart was moving left or right? Explain.

b. Are the force descriptions consistent with the motion description? For example, is the net horizontal force in the same direction as the acceleration? Give several specific examples.

c. At each position, compare the direction of the net force exerted by the spring on the cart and the cart's displacement from equilibrium when at that position.

19.2.2 Represent and reason

a. Construct five qualitative work–energy bar charts for the cart–spring system described in Activity 19.2.1 at the points described in the table that follows.

Construct a work–energy bar chart for point V.	Construct a work–energy bar chart for point IV.	Construct a work–energy bar chart for point III.
K U_s Other	K U_s Other	K U_s Other
+ 0 — — — −	+ 0 — — — −	+ 0 — — — −
Construct a work–energy bar chart for point II.	**Construct a work–energy bar chart for point I.**	
K U_s Other	K U_s Other	
+ 0 — — — −	+ 0 — — — −	

b. Do the charts depend on whether the cart is moving left when at a particular position or moving right? Explain.

c. How would the charts change if the surface had considerable friction? Explain.

19.2.3 Reason and explain Summarize the results of Activities 19.2.1–19.2.2 to describe and explain the motion of the cart. The description should include your observations, and the explanations should include reasoning based on force and energy analyses for the observed phenomena.

19.2.4 Represent and reason You have a small bob on a long string (a pendulum). The pendulum bob swings back and forth, as shown in the figure. At each of the marked points in the figure, the coordinate system consists of an axis in the radial direction (r axis) and a perpendicular axis in the tangential direction (t axis). Disregard air resistance.

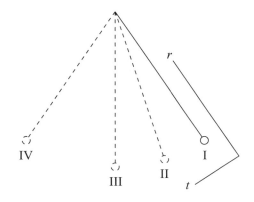

a. Complete the table that follows for positions shown in the figure.

Use the graphical velocity–subtraction method to estimate the bob's acceleration direction.	Position I	Position II	Position III	Position IV
Draw a force diagram for the bob.	Position I	Position II	Position III	Position IV
Draw the r component of the net force $\vec{F}_{net}$. Does the acceleration have a component in the radial direction that points in the same direction as the r component of the net force?	Position I	Position II	Position III	Position IV
Draw the t component of the net force $\vec{F}_{net}$. Does the acceleration have a component in the tangential direction that points in the same direction as the t component of the net force?	Position I	Position II	Position III	Position IV
Construct an energy bar chart.	Position I	Position II	Position III	Position IV

©2014 Pearson Education.

b. Is there a relationship between the *t* component of the net force and the displacement of the bob from the equilibrium position? Explain.

c. Compare the patterns of the net force and acceleration for the vibrating pendulum bob to the net force and acceleration of the vibrating cart in Activity 19.2.1.

19.3 Quantitative Concept Building and Testing

19.3.1 Observe and find a pattern Suppose that when the cart in Activity 19.2.1 was vibrating at the end of a spring, you used a motion detector to record the cart's motion. Graphs of position-versus-time, velocity-versus-time, and acceleration-versus-time are shown below.

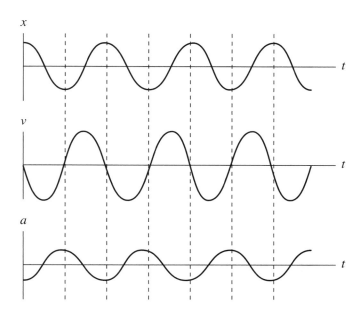

a. Examine the graphs carefully and answer the questions in the table that follows.

Recall that $v = \Delta x / \Delta t$. Is the shape of the velocity-versus-time graph consistent with this mathematical definition and with the position-versus-time graph? Compare the slope of $x(t)$ with the value of v at the maximum, minimum, and zero points. Explain.	Are these three graphs consistent with the force and motion diagrams in Activity 19.2.1? Explain.
Recall that $a = \Delta v / \Delta t$. Is the shape of the acceleration-versus-time graph consistent with this definition and with the velocity-versus-time graph? Compare the slopes of $v(t)$ with the value of a at the maximum, minimum, and zero points. Explain.	Are the direction and magnitude of the acceleration consistent with the direction and magnitude of the restoring force? Explain.

(continued)

Describe the relationship between the position-versus-time graph and the acceleration-versus-time graph. Explain why they mirror each other. Think about Newton's second law and the expression for the force that the spring exerts on the cart ($F_{S \text{ on } C_x} = -kx$).	Decide what mathematical function can be used to describe the position of the cart as a function of time.

b. Period is a physical quantity that characterizes the time interval for one complete vibration. Indicate in each of the three graphs at the beginning of this activity the period of the vibration.

19.3.2 Reason Assume that position, velocity, and acceleration change with time, as you saw in Activity 19.3.1. Write mathematical expressions for $x(t)$, $v(t)$, and $a(t)$ as cosine or sine functions of time. (In your expressions, try to use the quantities amplitude A and period T.) How do you know if the expressions you wrote make sense? *Hint:* Think about whether a sine or cosine function will work for the position-versus-time graph in Activity 19.3.1. Examine special cases (such as $t = 0$) and the relationships between the functions.

19.3.3 Observe and explain We would like to derive an approximate expression for the period of the cart's vibration at the end of the spring shown in Activity 19.2.1. Begin the derivation by answering two questions (a–b).

a. What physical quantities might affect the period?

b. Describe experiments that you could perform to decide if these quantities do in fact affect the period.

Now we move to the actual step-by-step derivation (c–g).

c. What is the total distance, in terms of the amplitude A, traveled by the cart during one complete cycle of vibration?

d. Assume that the cart's average speed v_{ave} during that cycle of vibration is half the cart's maximum speed v_{max}. Write an expression for the period T of the cart in terms of A and v_{max}.

e. Now use the fact that the maximum elastic energy of the spring equals the maximum kinetic energy of the cart to rewrite the expression for T in terms of the cart's mass m and the spring constant k.

f. A more rigorous method of derivation using calculus gives us a different expression for the period: $T = 2\pi (m/k)^{\frac{1}{2}}$. What is the difference between this expression and the one you provided in the previous part?

g. List all of the assumptions that you made to derive the expression for the period.

19.3.4 Test your ideas Assemble a set of springs, a set of bobs of different mass, and a ruler. Design an experiment to test each relationship proposed in the left column of the following table (some relationships may be incorrect). Fill in the table.

The period of vibration of a cart–spring system (as discussed in Activity 19.2.1) or of a pendulum bob depends on the amplitude of vibration.	Describe the experiment and include a sketch.	List the controlled variables (i.e., what you keep constant).	Describe the procedure and the predictions that you make based on the relationship you will test.	Perform the experiment and record the outcome.
The period of vibration of a pendulum bob depends on the mass of the bob.	Describe the experiment and include a sketch.	List the controlled variables (i.e., what you keep constant).	Describe the procedure and the predictions that you make based on the relationship you will test.	Perform the experiment and record the outcome.
The period of vibration of a system consisting of a bob attached to a vertical spring depends on the mass of the bob.	Describe the experiment and include a sketch.	List the controlled variables (i.e., what you keep constant).	Describe the procedure and the predictions that you make based on the relationship you will test.	Perform the experiment and record the outcome.

19.3.5 Observe and find a pattern You have a pendulum consisting of a long string with a metal bob hanging at one end. You can vary the mass m of the metal bob, the length L of the string, and the amplitude A of its back-and-forth vibration. You vary one of these three quantities at a time, measure the period of the pendulum, and then calculate the frequency (if needed). The data are shown in the table.

Bob mass (kg)	String length (m)	Amplitude (m)	Period (s)
1	1	0.05	2.00
2	1	0.05	2.00
3	1	0.05	2.00
1	2	0.05	2.80
1	3	0.05	3.45
1	4	0.05	4.00
1	1	0.07	2.00
1	1	0.10	2.00

a. Based on this data, decide what physical quantities affect the period. Decide which quantities do not affect the period. Explain.

b. An expression for the period of a *simple pendulum* (a small object vibrating with small amplitude on a long string) derived using calculus is:

$$T = 2\pi\sqrt{\frac{L}{g}}$$

where g is the acceleration due to gravity. Use the data in the table to decide whether the pendulum in the experiment can be considered a simple pendulum. Explain your decision.

19.3.6 Test your ideas Assemble a pendulum with a long string and a small metal bob. Use the relationship between the period and the length of the string for a simple pendulum to predict the period of its vibrations. Write the predicted value; take experimental uncertainties into account. Then perform the experiment and test your prediction. Record the experimental value and compare it to the prediction. Can your pendulum be considered a simple pendulum?

19.4 | Quantitative Reasoning

19.4.1 Represent and reason A 2.0-kg cart attached to a horizontal spring vibrates on a low-friction track (similar to the situation shown in Activity 19.2.1). The cart's displacement-versus-time is described by:

$$x = (0.20 \text{ m}) \sin\left[\left(\frac{2\pi}{2.0 \text{ s}}\right)t\right]$$

The positive direction of the x axis is to the right. Determine the cart's position at $t = 0$, $t = T/4$, $t = T/2$, and $t = (3T)/4$.

$t = 0$: _____ $t = T/4$: _____ $t = T/2$: _____ $t = (3T)/4$: _____

19.4.2 Represent and reason A 2.0-kg cart attached to a spring undergoes simple harmonic motion on a low-friction surface. Its displacement-versus-time is described by:

$$x = (0.20 \text{ m}) \sin\left[\left(\frac{2\pi}{2.0 \text{ s}}\right)t\right]$$

Complete the table that follows.

Determine the period and the amplitude of the motion.
Determine the spring constant.
Determine the maximum elastic potential energy of the system.
Determine the maximum speed of the cart.
Write an expression for the velocity as a function of time.

19.4.3 Represent and reason A 2.0-kg cart attached to a spring undergoes simple harmonic motion so that its displacement-versus-time is described by:

$$x = (0.20 \text{ m}) \sin\left[\left(\frac{2\pi}{2.0 \text{ s}}\right)t\right]$$

a. Draw a motion diagram for one cycle of the cart's motion.

b. Draw a force diagram for the cart at the times indicated in the table that follows.

$t = 0$	$t = T/4$	$t = T/2$	$t = 3T/4$	$t = T$

19.4.4 Represent and reason A 2.0-kg cart attached to a spring undergoes simple harmonic motion so that its displacement-versus-time is described by:

$$x = (0.20 \text{ m}) \sin\left[\left(\frac{2\pi}{2.0 \text{ s}}\right)t\right]$$

Complete the table that follows. Make sure the graphs are consistent with the motion diagram and the force diagrams in Activity 19.4.3.

Construct a position-versus-time graph.	*x* 0 ⊢——┼——┼——→ *t* 1 2
Construct a velocity-versus-time graph.	*a* 0 ⊢——┼——┼——→ *t* 1 2
Construct an acceleration-versus-time graph.	*v* 0 ⊢——┼——┼——→ *t* 1 2

19.4.5 Represent and reason A 2.0-kg cart attached to a spring undergoes simple harmonic motion so that its displacement-versus-time is described by:

$$x = (0.20 \text{ m}) \sin\left[\left(\frac{2\pi}{2.0 \text{ s}}\right)t\right]$$

Construct qualitative energy bar charts for the cart–spring system at the times indicated in the table that follows.

$t = 0$	$t = T/4$	$t = T/2$
K U_s Other	K U_s Other	K U_s Other
+ 0 ─ ─ ─ −	+ 0 ─ ─ ─ −	+ 0 ─ ─ ─ −

$t = 3T/4$	$t = T$	
K U_s Other	K U_s Other	
+ 0 ─ ─ ─ −	+ 0 ─ ─ ─ −	

19.4.6 Regular problem An astronaut living at a space station is in a constant state of free fall—the only force exerted on her and on the space station is the gravitational force due to Earth. How can an astronaut determine her mass while on an extended stay in space? One method involves vibrational motion. An astronaut sits on a chair that vibrates horizontally at the end of a spring. A motion detector determines the amplitude of vibration and the speed of the chair as it passes through the equilibrium position. The spring has a 1200-N/m spring constant, the amplitude of vibration is 0.50 m, and the speed of the chair as it passes through equilibrium is 2.0 m/s. Using these data, find the combined mass of the chair and the astronaut.

19.4.7 Design an experiment Devise another method to determine the mass of the astronaut described in Activity 19.4.6. You know the spring constant of the spring–seat system (1200 N/m) and the mass of the vibrating seat on which she sits. You also have a stopwatch. Describe the method you will use, and give a sample calculation.

19.4.8 Regular problem A 0.20-kg arrow moving horizontally at 10 m/s hits a 0.40-kg clay ball hanging at the end of a 1.5-m-long string. The arrow sticks in the clay ball, and the arrow and ball swing together in an arc up to some undetermined final height.

a. Determine the speed of the ball and arrow system immediately after the collision.

b. The ball with the arrow swings in an arc upward after the collision. Determine how high the ball will rise and whether the answer is reasonable.

c. Determine the frequency of the pendulumlike vibration of the ball–arrow system.

d. List the assumptions that you made.

19.4.9 Equation Jeopardy Mathematical expressions describe two situations involving vibrational motion. Fill in the table that follows. Provide all the details for these situations. *Note:* There is more than one possible solution for each problem.

Mathematical description	Sketch a situation the equation(s) might describe.	Write in words a problem for which the equation(s) is a solution.
$(1/2)(20{,}000 \text{ N/m})(0.20 \text{ m})^2$ $= (1/2)(100 \text{ kg})v^2 + (1/2)(20{,}000 \text{ N/m})(0.10 \text{ m})^2$		
$0.20 \text{ Hz} = (1/2\pi)[k/(100 \text{ kg})]^{1/2}$ $(1/2)k(0.40 \text{ m})^2 = (1/2)(100 \text{ kg})v_{max}^2$		

19.4.10 Evaluate the solution *The problem:* You are helping design a stopping system for Soapbox Derby race cars. The proposed stopper is a padded cushion that catches the car at the end of the race. The far end of the cushion, opposite the car, is attached to a spring that compresses at collision. The car's mass with driver is 60 kg, and it is traveling at 12 m/s when it first contacts the stopper; the mass of the stopper is 20 kg. The car is to stop in 2.0 m after contacting the stopper. What is the spring constant of the spring needed for this system?

Proposed solution: The initial kinetic energy of the car is converted into elastic potential energy of the compressed spring when the car stops. Thus,

$$(1/2)mv^2 = (1/2)kA^2$$

or

$$k = m(v^2/A^2)$$
$$= (60 \text{ kg})[(12 \text{ m/s})^2/(2.0 \text{ m})^2] = 2160 \text{ kg/s}^2$$

a. Identify any errors in the solution to this problem.

b. Provide a corrected solution if there are errors.

19.4.11 Design an experiment You have a stop watch and a metal ball attached to a 1.0-m-long string. Design an experiment using this equipment to measure the acceleration of free fall g. Fill in the table that follows to help you.

Describe the experiment in words.	List the physical quantities that you will measure and the quantities that you will calculate.
	To be measured:
Draw a labeled sketch of the apparatus.	*To be calculated:*
Describe the mathematical procedure you will use to determine g.	List sources of experimental uncertainties and ways to minimize them.
	Uncertainties:
List additional assumptions.	
Perform the experiment and record the outcome. Does your result make sense?	*Ways to minimize:*

19.4.12 Design an experiment Design an experiment to determine if a human arm can be treated as a simple pendulum when it swings back and forth during a walk. Fill in the table that follows.

Describe the experiment in words.	List the physical quantities that you will measure and the quantities you will calculate.
	To be measured:
Draw a labeled sketch of the apparatus.	*To be calculated:*
Describe the mathematical procedure you will use.	List sources of experimental uncertainties and ways to minimize them.
List additional assumptions.	*Uncertainties:*
Describe how you will make a judgment about whether an arm can be treated as a simple pendulum.	*Ways to minimize:*

19.4.13 Pose a problem You have a rubber band, a 100-g object, a stopwatch, and a meterstick. Pose an experimental problem that you can solve using this equipment. Describe how you would solve the problem.

20 Mechanical Waves

20.1 | Qualitative Concept Building and Testing

20.1.1 Observe and find a pattern Fasten one end of a metal Slinky® toy to the leg of a chair resting on a hard floor—or alternatively, to a clamp, which in turn is fastened to the end of a lab bench so the Slinky lies on the smooth lab bench surface. Grasp the free end of the Slinky and stretch it so that the Slinky is about 3- to 4-m long. Do not lift it off the smooth surface. Fill in the table that follows.

a. Keeping the Slinky on the smooth surface and stretched along a straight line, give the end of the Slinky in your hand a quick push along its axis. Describe what you observe.	
b. Sketch the Slinky at one instant of time during the propagation of the disturbance you created in part a.	
c. Indicate in words or draw how an individual Slinky ring in the middle of the Slinky moves with respect to the Slinky as the disturbance passes.	
d. If you repeat the procedure described in part a, but this time push more abruptly, does the disturbance move faster along the Slinky? How do you know?	
e. If you push less abruptly than in part a, does the speed of the disturbance change? How do you know?	

Slinky® is a registered trademark of Poof-Slinky Inc.

20.1.2 Observe and find a pattern Keep the Slinky toy from Activity 20.1.1 on a smooth, hard surface and fastened securely at one end. Again grasp the free end of the Slinky and stretch it so that the Slinky is about 3- to 4-m long. Fill in the table that follows.

a. Give the end of the Slinky in your hand an abrupt sideways shake, perpendicular to the Slinky (see the top-view illustration), all the while keeping it on the smooth surface. Describe what you observe. Top view 🖐〰〰〰〰〰〰〰◆ ↓	
b. Sketch the Slinky at one instant of time during the propagation of the disturbance you created in part a.	
c. Indicate in words or draw how an individual Slinky ring in the middle of the Slinky moves with respect to the Slinky as the disturbance passes.	
d. If you repeat the procedure described in part a, but this time make a larger abrupt sideways shake, does the disturbance move faster along the Slinky?	
e. If you make a smaller abrupt sideways shake than in part a, does the speed of the disturbance change? How do you know?	

20.1.3 Observe and find a pattern Suppose you stand in the water a meter or two from one end of a swimming pool.

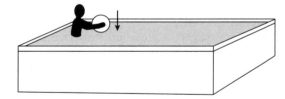

a. You push down hard once on a large beach ball floating on the water in front of you. What do you observe?

b. How can you estimate the speed at which the disturbance moves away from the ball?

c. Now you push down on the ball a few times (do not push too many times). Does the speed at which the crests of the waves move away from you depend on how frequently you push the ball up and down? Explain.

d. Draw a top view of the wave-crest pattern that you would see in the water at one instant of time. How does the distance between adjacent crests differ if you bob the ball up and down more frequently or less frequently? Explain.

20.1.4 Reason and explain Small balls of mass m are connected with small springs of spring constant k, as shown in the figure. The balls and springs rest on a smooth, frictionless surface. Imagine that you vibrate one end of this chain of balls and springs back and forth, parallel to the axis of the chain, causing a wave disturbance that moves along the spring–mass chain at some speed v.

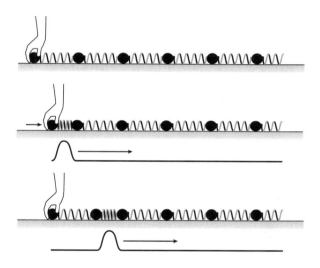

a. Do you think the speed of the waves along the chain depends on the spring constant k? If so, do you think the speed is greater or less for greater spring constants? Explain.

b. Do you think the speed of the waves along the chain depends on the mass m of the balls? If so, do you think the speed is greater or less for greater mass? Explain.

c. By analogy, identify two properties of stretched strings (for example, a violin, guitar, or piano string) that might affect the speed of waves on the strings.

20.1.5 Design an experiment Design an experiment to determine if a transverse pulse or a longitudinal pulse moves with greater speed along a Slinky.

Sketch the experimental setup.	Explain in words how you will measure the speed for each kind of pulse.	List quantities that you will measure and quantities that you will calculate. List your assumptions and experimental uncertainties.	Perform the experiment; record the results and describe your conclusion.
		To be measured: To be calculated: Assumptions: Uncertainties:	

20.1.6 Observe and explain Each of the two waves depicted in the table represents a disturbance of a medium at one instant of time. The disturbance travels away from the source of the disturbance. For each wave, construct a graph that shows along the vertical axis how the medium is disturbed at that instant of time at different positions x along the horizontal axis. Be sure to label the vertical axes with the names of the quantities that you are plotting on those axes.

Wave on string	Sound wave in air
Picture of the disturbance at one instant in time	Picture of the disturbance at one instant in time
Graphical representation of the disturbance	Graphical representation of the disturbance

20.1.7 Observe and explain Consider a beach ball bobbing up and down in the center of a swimming pool. Imagine that the ball remains at the same position at the center of the pool. The illustration shows several consecutive wave crests at one instant of time. Suppose that observer *A* is moving toward the source and observer *B* is moving away from the source. Does *A* or *B* observe higher frequency water wave vibrations, or do they observe the same frequency vibrations? Explain.

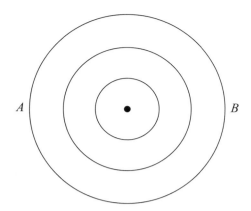

20.1.8 Observe and explain Four wave pulses produced by a large beach ball bobbing up and down in a pool are shown in the figure as the ball moves to the right. Wave-crest 1 of large radius was created when the ball was at position 1, and the wave-crest 4 of small radius was created when it was at position 4. Explain why observer *B*, standing stationary in the water in the direction the source moves, feels higher frequency water wave pulses than observer *A*, standing stationary behind the moving wave source.

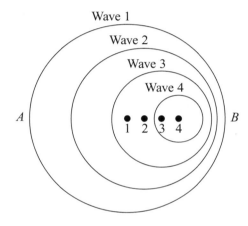

20.1.9 Describe and explain Look for a pattern in the two experiments described next.

Experiment I: You stand beside a train track; a train blowing its whistle moves toward you, passes you, and then moves away. The sound from the whistle changes from a higher pitch (higher frequency) as it moves toward you to a lower pitch as it moves away.

Experiment II: Your professor swings a ball tied to a rope in a horizontal circular path. A whistle inside the ball makes a higher-pitched sound as the ball moves toward you and a lower-pitched sound as it moves away. Devise a qualitative explanation for these observations using the ideas you developed in Activity 20.1.8.

20.2 | Conceptual Reasoning

20.2.1 Represent and reason A longitudinal wave of amplitude 3.0 cm, frequency 2.0 Hz, and speed 3.0 m/s travels on an infinitely long Slinky. Displacement y is the distance that some part of the Slinky is displaced from its equilibrium position.

a. How far apart are the two nearest points on the Slinky that at one particular time both have the maximum displacements from their equilibrium positions? Explain your reasoning.

b. Complete the graphs below. Position x is a point along the axis of the Slinky. Be sure to put scales on the graphs.

Construct a displacement-versus-time graph for one coil of the Slinky. Show the period T of the wave on the graph.	Construct a displacement-versus-position graph for a segment of the infinitely long Slinky. Show the wavelength of the wave on the graph.
y ... t	y ... x

20.2.2 Reason The frequency f of a wave equals $1/T$.

a. Explain why this makes sense.

b. Suppose that there are 10 vibrations in 5 s. What is the frequency of such a wave, and what is its period?

c. If the wave travels at speed 4.0 m/s, how far will it travel during one period?

d. Show that $\lambda = v/f$.

20.2.3 Reason The speed of a wave depends on properties of the medium through which the wave travels. The speed v can also be determined in a different way—if you know the wavelength λ of the wave and the period T of the vibration, then $v = \lambda/T$.

a. Explain why this equation makes sense.

b. Is the equation $v = \lambda/T$ an operational definition or a cause-effect relationship for the wave speed? Explain.

20.2.4 Reason The graph describes the varying air pressure against a microphone as a sound wave passes. The readings are given with respect to the atmospheric pressure when the sound is not present. The negative pressure means that the pressure at a particular time is less than atmospheric pressure. *Note:* Sound travels at about 340 m/s. Note that 1 Pa (pascal) $= 1\ \text{N/m}^2$.

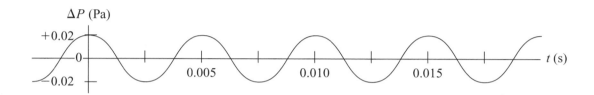

a. Determine the amplitude of the wave.

b. Determine the frequency of the wave.

c. Determine the wavelength of the wave.

20.2.5 Reason The graph describes the varying air pressure at different positions in space at one particular time due to a sound wave. Again, remember that sound travels at about 340 m/s.

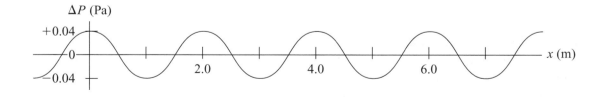

a. Determine the amplitude of the wave.

b. Determine the wavelength of the wave.

c. Determine the frequency of the wave.

20.2.6 Represent and reason Two waves are shown in the illustration that represent pressure variation-versus-time; they could, for example, be the pressure variations caused by two different sound waves at a microphone.

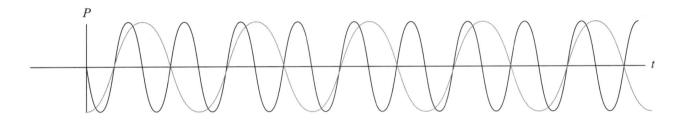

Construct the complex wave that would be produced if both waves were present at the same location during that same time interval.

20.3 | Quantitative Concept Building and Testing

20.3.1 Describe and explain A periodic wave disturbance at one particular time (call it $t = 0$) is represented by the graph. In a way the graph is a snapshot of the wave. Your physics major friend claims that the equation that follows describes this periodic wave disturbance at different positions x at different times t:

$$y = A \cos 2\pi (t/T - x/\lambda)$$

where y is the disturbance at time t of the medium at position x (the distance from the source). Answer the following questions to try to disprove or support her claim. Note that A is the amplitude of the wave, T is its period and λ is its wavelength.

a. At $t = 0$ and $x = 0$, this particular wave disturbance has a value $y = A \cos 0 = A$, which matches what we see in the figure. At the same time ($t = 0$), what will the value y of the disturbance be at one wavelength forward? What will it be at two wavelengths forward? Three wavelengths forward? Does the equation give you the desired value? Explain.

b. At $t = 0$, what will the value y of the wave disturbance be at positions $x = \lambda/2, 3\lambda/2, 5\lambda/2$, and so forth? Does the equation give you the desired values? Explain.

c. At $t = 0$, what will the value y of the wave disturbance be at positions $x = \lambda/4, 3\lambda/4, 5\lambda/4$, and so forth? Does the equation give you the desired value? Explain.

d. At $t = T$, what will the value of y be at positions $x = 0, \lambda/4, \lambda/2, 3\lambda/4$, and λ? Does the equation give you the desired values? Explain.

e. Does the mathematical description seem appropriate based on your analysis? Explain.

20.3.2 Reason The equations that follow describe the variation of pressure at different positions and times (relative to atmospheric pressure) caused by sound waves. Fill in the table that follows. *Note:* The speed of sound in air is 340 m/s.

Equation $\Delta P = (2.0 \text{ N/m}^2) \cos 2\pi[t/(0.010\text{ s}) - x/(3.4\text{ m})]$			
Identify the amplitude of the pressure variation.	Identify the period for one vibration.	Determine the frequency of the sound.	Identify the wavelength of the sound.

Equation $\Delta P = (4.0 \text{ N/m}^2) \cos 2\pi[t/(0.0010\text{ s}) - x/(0.34\text{ m})]$			
Identify the amplitude of the pressure variation.	Identify the period for one vibration.	Determine the frequency of the sound.	Identify the wavelength of the sound.

20.3.3 Observe and explain Imagine that you have three long springs. If you measure the speed of wave pulses along the springs, you would accumulate the data given in the table.

Spring number	Force exerted on the end of the spring (tension, N)	Amplitude (cm)	Frequency (Hz)	Mass/length (kg/m)	Speed (m/s)
1	4.0	10	2	0.16	5.0
1	8.0	10	2	0.16	7.1
1	16.0	10	2	0.16	10.0
1	4.0	10	2	0.16	5.0
1	4.0	20	2	0.16	5.0
1	4.0	30	2	0.16	5.0
1	4.0	10	2	0.16	5.0
1	4.0	10	3	0.16	5.0
1	4.0	10	4	0.16	5.0
1	4.0	10	2	0.16	5.0
2	4.0	10	2	0.080	7.1
3	4.0	10	2	0.040	10.0

Come up with an expression that can be used to determine the speed of the wave as a function of different properties of the springs.

20.3.4 Describe

a. Write an equation that describes the graph line shown at the right. Note that v is the speed of a wave on a string, F is the magnitude of the force exerted on the end of the string (equal to the tension in the string), and μ is the mass per unit length of the string. Is this equation consistent with the data in Activity 20.3.3?

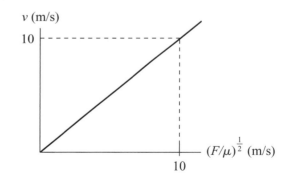

b. Determine the speed of a wave pulse on a 0.080-kg/m string when a person pulls exerting a 2.0-N force on the end of the spring.

20.3.5 Observe and find a pattern Tie one end of a rope (or one end of a long, tightly wound spring) securely to a post. Hold the other end in your hand and vibrate it up and down at different frequencies. At most frequencies, the rope responds little to your efforts. However, at special frequencies, big amplitude vibrations occur (the figure that follows shows four of these vibrations). If you videotape the process and view the video frame by frame, you will find that the period of the second vibration is half the period of the first, the period of the third is one-third of the period of the first, and so forth.

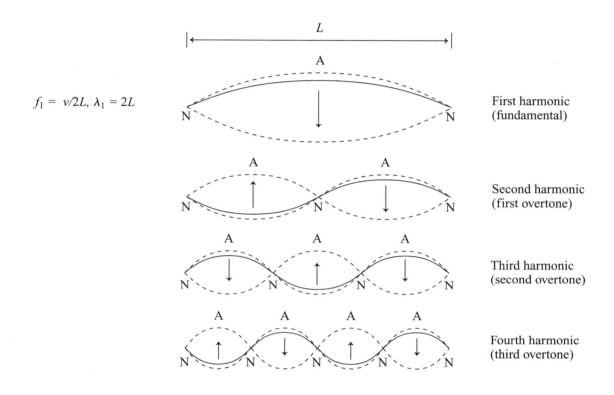

$f_1 = v/2L$, $\lambda_1 = 2L$

First harmonic (fundamental)

Second harmonic (first overtone)

Third harmonic (second overtone)

Fourth harmonic (third overtone)

a. Write expressions for the frequency of the other standing-wave vibrations; there are many more possible than are shown in the figure.

b. Write expressions for the wavelengths of the observed and of other standing-wave vibrations.

20.3.6 Test your idea Pluck the A string on a violin (or another stringed instrument). Now tune an electronic oscillator attached to a sound speaker, and you will find that this string vibrates at 440 Hz (called concert A). The string is 0.33 m long.

a. Find the speed of a pulse on the string.

b. Use the information in part a and what you learned in Activity 20.3.5 to predict the frequency of the string's vibration if you press your finger against the fingerboard, thus effectively changing the length of the string to 0.22 m. You then pluck it again.

c. Check your prediction by plucking the string and matching the sound to that of the oscillator. Did your prediction match the outcome? (You can try the same type of experiment with other stringed instruments.)

20.3.7 Explain Wind instruments such as trumpets, flutes, clarinets, and the pipes in organs consist of columns of air inside tubes. They also have opening and closing valves (or slides, in the case of a trombone).

a. Explain how the valves and slides allow a musician to change the frequency of sound that these instruments produce.

b. Blow a whistle and try to change the frequency of sound it produces. How did you do it? Is the reason consistent with the explanation you provided in part a?

20.3.8 Test your idea We express the fundamental frequencies of vibration of a tube open at both ends (called an open tube) the same as the expression for a string $(f_n = nv/2L)$, where v is the speed of sound in the air inside the tube, L is the length of the tube, and n is an integer $(1, 2, 3 \dots)$. You have a whirly tube that is 0.86 m long and is open at both ends. When swung in the air, it produces a sound of a particular frequency. If swung faster, the frequency is higher. You can get about three distinct frequency sounds from a whirly tube.

a. Use the expression for the standing-wave frequencies in tubes to predict the frequencies of the whirly tube.

b. Check your predictions by whirling the tube while simultaneously tuning an electronic oscillator connected to a sound speaker to get matching frequencies. Compare what you measure with your predictions.

c. Do your predictions agree (with minor difference) with the measured value? If not, discuss the observed difference.

20.3.9 Reason and explain

a. We found in Activity 20.1.7 that the observed frequency is higher if the observer moves toward the stationary source and is lower if the observer moves away from the stationary source. Use the equation given earlier and sign conventions to show that these changes in frequency are consistent with the equation.

b. We observed and predicted in Activity 20.1.8 that the observed frequency is higher if the source moves toward the stationary observer and is lower if the source moves away from the stationary observer. Show that these changes in frequency are consistent with the equation.

20.4 ∎ Quantitative Reasoning

20.4.1 Represent and reason A traveling wave is represented in different ways in the table that follows.

Description in words	A tuning fork of frequency 100 Hz produces sound in the air. The sound travels at speed 340 m/s and causes a maximum pressure variation of 0.020 N/m². The wave is represented along one direction in other ways.
Graph: **Pressure variation-versus-time for one position**	
Mathematics:	$T = 0.01$ s
Graph: **Pressure variation-versus-position graph at one instant in time**	
Mathematics:	$\lambda = 3.4$ m
Mathematics: **Equation of displacement as a function of position and time**	$\Delta P = (0.020 \text{ N/m}^2) \cos 2\pi [t/(0.010 \text{ s}) - x/(3.4 \text{ m})]$

Are the representations consistent with each other? Explain.

20.4.2 Represent and reason Fill in the table that follows to describe in multiple ways a traveling wave on a very long string. The period of the wave is 0.20 s, the amplitude is 6.0 cm, and the wavelength is 6.0 cm. Make sure you use consistent units.

Construct a displacement-versus-time graph for one point on the string.		Determine the wave's frequency and speed.	
Construct a displacement-versus-position graph for one particular time.		Write an equation for the displacement as a function of position and time.	

20.4.3 Represent and reason Two ropes are the same length. The speed of a pulse on rope 1 is 1.4 times the speed on rope 2.

Write an expression for the ratio of the speeds (v_1/v_2) in terms of the ratios of the rope tensions (F_1/F_2) and of the rope masses (m_1/m_2). Use no numbers yet.	If the forces pulling on the ends of the rope (rope tensions) are the same, determine the ratio of their masses for the speed ratio given in the problem statement.	If the masses of the ropes are the same, determine the ratio of the forces pulling on the ends of the ropes—for the speed ratio given in the problem statement.

20.4.4 Represent and reason Read the descriptions of the situations described below and answer the questions that follow.

Situation I: A 0.50-m-long string vibrates in three segments, with a frequency of 240 Hz.

Situation II: A 0.68-m pipe is open at both ends. The speed of sound is 340 m/s.

a. What is the fundamental frequency of the string?

b. What is the speed of a wave on this string?

c. What is the fundamental frequency of the pipe?

d. What is the fundamental frequency of the pipe when one end is closed?

20.4.5 Evaluate the solution A friend proposes a solution for the following problem.

The problem: A violin A string is 0.33 m long and has mass 0.30×10^{-3} kg. It vibrates at a fundamental frequency of 440 Hz (concert A). What is the tension in the string?

Proposed solution: Speed depends on the tension and string mass ($v = [T/m]^{\frac{1}{2}}$). Thus,

$$T = v^2m = (340 \text{ m/s})^2(0.30 \text{ g}) = 34{,}680 \text{ N}$$

a. Evaluate the solution and identify any errors.

b. Provide a corrected solution if you find errors.

20.4.6 Evaluate the solution

The problem: A shepherd blows on the end of a bone pipe (it is considered closed at one end and open at the other) that is 0.30 m long. She can play the first harmonic by blowing gently and higher harmonics by blowing harder. Determine the frequencies of these first three harmonics.

Proposed solution: The speed of sound in the solid bone material is about 3000 m/s and in air is about 340 m/s. Thus, the first three harmonic frequencies are:

$$f_1 = v/2L = (3000 \text{ m/s})/[2(0.30 \text{ m})] = 5000 \text{ Hz}$$
$$f_2 = 2v/2L = 2(3000 \text{ m/s})/[2(0.30 \text{ m})] = 10{,}000 \text{ Hz}$$
$$f_3 = 3v/2L = 3(3000 \text{ m/s})/[2(0.30 \text{ m})] = 15{,}000 \text{ Hz}$$

a. Identify any errors in the solution to the problem.

b. Provide a corrected solution if there are errors.

20.4.7 Regular problem You form a jug-and-bottle band where the musicians blow across the tops of the bottles (and the jug) to initiate sounds. Your band wants to play songs with notes ranging in frequency from 120 Hz to 300 Hz.

Determine the size of the jug or bottle to play the 120-Hz sound. Explain how you made your choice.		Determine the size of the jug or bottle to play the 300-Hz sound.	

20.4.8 Represent and reason How can you distinguish the sound of a violin and a flute both playing the same note—for example, concert A at 440 Hz? Most sounds are made up of a combination of the fundamental frequency of the vibrating object and a combination of higher harmonics. The quality of the sound depends in part on the number and relative amplitudes of these higher harmonics compared to the amplitude of the fundamental frequency. A rich violin sound may include 20 harmonics of A, and a flute sound may include only a few. The graph above represents the pressure disturbance-versus-position at one time of a sound wave of frequency 100 Hz.

a. Draw another wave of frequency 200 Hz with half the amplitude.

b. Add the two harmonics together to construct a so-called complex wave made of the two harmonic waves.

20.4.9 Regular problem The Doppler effect can be used to determine the speed of red blood cells (as well as baseballs and cars). The Doppler speed detector emits sound at a particular frequency and detects the reflected sound at a different frequency. The difference in the emitted and detected sound frequencies indicates the speed of the object being measured. Assume that sound of frequency 100,000 Hz enters an artery opposite the direction of blood flow, which travels at speed 0.40 m/s. Answer the questions below to see how detecting the frequency of the sound reflected from a red blood cell indicates how fast it is moving.

a. Use the Doppler equation to determine the frequency that the cell would detect as it moves toward the sound source.

b. Suppose that the moving cell emits sound at the same frequency it detected in part a. What frequency does the Doppler detection system measure coming from the cell?

c. Often the Doppler detection system measures a beat frequency. The beat frequency is the magnitude of the difference between the emitted sound and the reflected sound that it received back from the moving blood cell. What beat frequency is observed in the case described above?

21 Reflection and Refraction

21.1 | Qualitative Concept Building and Testing

21.1.1 Observe and explain Go to a room that is isolated from all external light sources—natural and artificial. Turn off the internal lights and wait in the dark room for several minutes. Record your observations and propose an explanation.

21.1.2 Observe and explain Place a laser pointer on a horizontal surface (say, a desk) in the center of a room and observe a bright spot on the wall toward which the laser points. Fill in the table that follows.

What path did light follow to reach the wall? You can find it by trial and error—by trying to block the light with a small piece of paper at several locations along its path to the wall.	
What can you say about the path of the light from the laser to the wall? Represent that light path by a long arrow, called a *ray*. A ray is not real; it is just a way to show the direction that light is traveling.	
Why can't you see the beam of light itself but you can see the bright spot on the wall or on a piece of paper that intersects with the beam? Write possible explanations.	
Now sprinkle chalk dust along the line of light propagation; you will see the beam of light in the air. Explain why the chalk dust made it possible to see the light beam.	
Discuss the conditions needed for us to see something.	

21.1.3 Test an idea Place a powered, frosted lightbulb on a table in the center of a dark room and observe that the walls are almost uniformly lighted. A friend draws two ray diagrams to try to explain this observation.

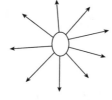
a.

b.

a. Describe the main difference between the two diagrams. Consider how each point of the bulb emits light according to the diagrams.

b. Design an experiment to test which of the diagrams represents the way a lightbulb emits light—does each point emit one ray, or does each point emit rays in all directions? Describe the experiment with a picture and write a prediction based on each diagram.

c. Perform the experiment and decide which diagram did not predict the outcome. Which diagram will you use to represent how each point of the bulb emits light?

21.1.4 Observe and explain Place a powered, frosted lightbulb on a table in the center of a dark room. Take the point of a sharpened pencil and place it *close* to the wall. Record what you observe. Next move the pencil toward the bulb and away from the wall.

a. Represent what you observe with words, a sketch, and a ray diagram.

b. Discuss how your ray diagram uses ideas from Activity 21.1.3.

21.1.5 Observe and explain Place a candle on a table and set a piece of thin cardboard with an opening cut into it on the table between the candle and a nearby wall (see the figure). Move the cardboard closer to the wall and then move it closer to the candle. Observe the changes on the wall.

  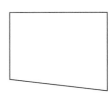

Candle   Cardboard   Wall

Positions of the cardboard	Draw ray diagrams to represent what you observe on the wall.
Cardboard close to the wall	
Cardboard in an intermediate position between the candle and the wall	
Cardboard close to the candle	

21.1.6 Test your ideas Imagine that you put a candle on a table and place a piece of thin cardboard between the candle and a nearby wall. Use the figure to draw a ray diagram to predict what you will see on the wall if you make a tiny hole in the cardboard. Then light the candle and turn off the room lights to observe the outcome of the experiment and revise your diagram if necessary.

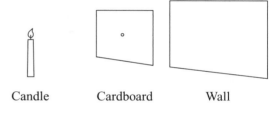

Candle   Cardboard   Wall

21.1.7 Observe and find a pattern Place a protractor on a tabletop and put the flat edge of the protractor against a mirror that is held upright and perpendicular to the tabletop and protractor. Shine a laser pointer across the protractor so that the beam hits slightly above the center zero point on the protractor. The reflected light returns across the protractor (see the figure). Then change the angle at which the laser ray hits the mirror and see if there is any pattern in the direction of the reflected light. Record your results in the table. Notice the normal line CO in the figure—this is a line perpendicular to the surface of the mirror at a point where the incident beam hits the mirror.

Top view

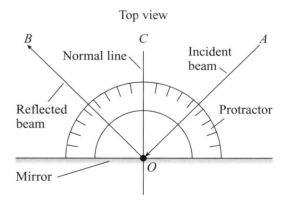

Angle AOB between the incident and reflected beams	Angle between the incident beam AO and the mirror	Angle between the reflected beam BO and the mirror	Angle between the incident beam AO and the normal line CO	Angle between the reflected beam BO and the normal line CO
	90°			
	70°			
	60°			
	45°			
	30°			
	10°			

Find a pattern in the data and express the pattern in words and mathematically.

21.1.8 Test your idea Assemble two mirrors on a flat surface so that their faces make a right angle, as shown in the figure. Place a target on the wall or on the other side of the table. Use any of the relationships that you found in Activity 21.1.7 to predict how you need to aim a laser beam to hit mirror 1 so that light reflected from mirror 1 then hits mirror 2 and finally hits the target.

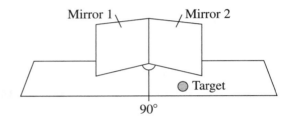

a. Draw a ray diagram to make a prediction.

b. Perform the experiment, record the results, and check to see if your prediction matches the outcome of the experiment.

c. Can you hit the target with the light that hits mirror 1 from a different direction? Explain.

21.1.9 Observe and find a pattern Shine the light from a laser beam on to the surface of a clear plastic container filled with water. The path of the light is shown in the illustration for three different incident angles of the laser beam.

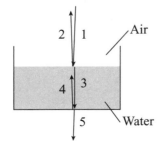 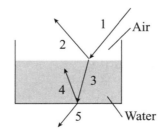 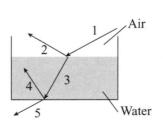

a. What happens to the beam of light that is incident on an interface between two different media—for example, ray 1 reaching the top surface of the water or ray 3 reaching the bottom? Traditionally physicists use a line perpendicular to the air–water interface at the point at which a ray strikes it to record the changes in the direction of the beam.

b. For the three situations illustrated, describe any pattern(s) you observe when looking at rays 1 and 2 or rays 3 and 4.

c. For the three situations illustrated, describe any pattern(s) you observe when looking at rays 1 and 3 or rays 3 and 5.

d. For the three situations illustrated, describe any pattern(s) you observe when looking at rays 1 and 5. Is this pattern consistent with the patterns you discussed in part c? Explain.

21.1.10 Explain In the previous activities you found that light reflects and refracts (bends) as it travels between different media. One explanation scientists formulated years ago to explain this observed phenomena was a "particle model" of light. They thought that an object that radiates light emits tiny particles, like little bullets, that travel in all directions. Use this model to explain the following.

a. How does light travel in straight lines in the same medium?

b. How does light form shadows if it encounters obstacles?

c. How does the angle of incidence equal the angle of reflection?

d. How does light bend when passing from one medium into another, different medium?

e. Light bends toward the normal line when it travels from air into any other medium. What do you need to assume about the components of velocity parallel to the surface and perpendicular to the surface of a light particle as it passes from air to the second medium?

21.1.11 Represent and reason Christiaan Huygens, a contemporary of Newton, developed a wave model of light that competed with Newton's (at the time) more popular particle model. Huygens wondered what would happen if several waves simultaneously traveled through a medium. To answer this question, let's try a paper-and-pencil experiment similar to that used by Huygens. We mark six dots across a page, each dot separated by 1 cm from the adjacent dot (see the illustration). The dots represent points on the crest of a wave moving toward the top of the page. According to Huygens, each dot is the source of a small wave disturbance that moves up the page in the direction the wave is traveling. In the figure the 3-cm-radius half circles, called *wavelets,* represent these disturbances. On the illustration provided, note places above the dots where the net disturbance from the six wavelets is two or more times bigger than the disturbance caused by any one wavelet—places where the wavelets add together to form bigger waves. This is the new crest of the wave. Draw a line on the sketch indicating the location of the new wave crest that was formerly at the position of the dots. Also, draw a ray indicating the direction the wave is traveling. The pattern is even clearer if you make many more dots and wavelets in the same space.

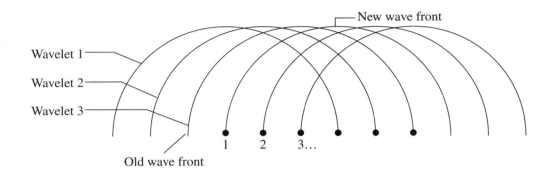

21.1.12 Represent and reason In Activity 21.1.11, semicircular wavelets all traveling at the same speed make up a new wave front. Waves often travel at different speeds in different places. For example, water waves travel slower in shallow water than in deeper water. Sound travels slower in cold air than in warm air. The difference in speed in different regions causes the wave to bend—to change direction. Huygens' principle can be used to understand this better. In the sketch below, the six dots are part of a wave crest (a wave front) moving toward the top of the page. The wave travels slower on the left side than on the right side. Thus, the wavelets originating from the left side have smaller radii than those farther to the right.

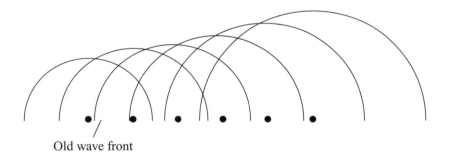

Old wave front

a. On the sketch, draw a new wave front that is produced by the wavelets leaving the positions of the six dots—that is, leaving the old wave front. Also, draw a ray that approximately indicates the wave's path as it moves up the page.

©2014 Pearson Education.

b. Based on this activity, which way do you think that waves tend to bend—toward regions in which they travel slower (the left side of the page) or regions in which they travel faster (the right side of the page)? Explain. (Remember your answer; you'll use this idea later.)

c. Based on your answer to part b and looking at the sketch in Activity 21.1.9, decide if light seems to travels faster or slower in water than in air. Explain your answer.

21.1.13 Represent and reason Imagine a wave whose wave fronts moving in one medium are incident on a boundary with another medium. The wave ray is not perpendicular to the boundary of the two media (see the illustration). The wave travels faster in medium 1 than in medium 2. During a certain time interval, the wavelet that earlier left the right edge of the wave front is just reaching the boundary between medium 1 and medium 2. The wavelet that left the lower left edge of the wave front at the same time travels less distance (i.e., moves more slowly) in medium 2.

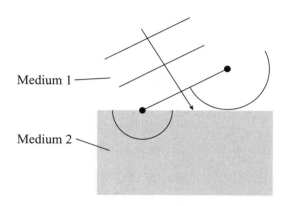

a. The wavelets leaving the middle of the wave front travel part of the time in faster medium 1 and part of the time in slower medium 2. Note that their radii should be longer than the wavelet on the lower-left edge but shorter than the wavelet on the upper-right edge. What is the orientation of the new wave front formed from these wavelets—now completely in medium 2? Draw a ray indicating the direction of the wave in medium 2.

b. Compare your sketch with the illustration in the Activity 21.1.9. Based on your analysis here and on that sketch, decide if light travels faster or slower in water than in air. Explain your answer.

21.2 | Conceptual Reasoning

21.2.1 Represent and reason Draw ray diagrams to explain why the shadow of your standing body gets shorter as the Sun rises higher above the horizon. Assume that the Sun is infinitely far away and Earth receives only parallel rays of light.

21.2.2 Represent and reason Draw a ray diagram to determine the angle of the Sun relative to the horizon when the shadow of your body is the same length as your body.

21.2.3 Represent and reason Parts a and b of the figure below depict the path of light beams moving from air into water (the container holding water is made of glass with very thin walls, so that the boundary air/glass and water/glass can be disregarded) or glass and then out into air again. (The reflected beams are not shown.) Note the bending of the light at the interfaces between the two media. In particular, keep track of the light path relative to the normal lines that are perpendicular to the interfaces. Based on the patterns you observe in parts a and b, draw onto the figure for part c a normal line and a ray indicating the light path after it moves from the water to the air. For part d, draw normal lines and rays as the light moves from the air into the glass, through the glass, and out into the air on the other side.

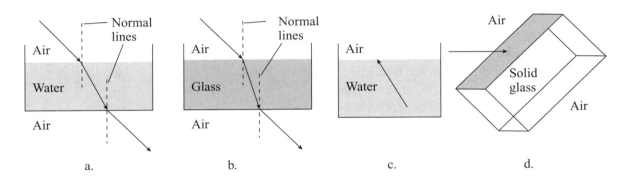

a.   b.   c.   d.

21.2.4 Represent and reason Imagine that you place a closed, empty glass box (with thin glass walls) filled with air underwater and hold it there. A beam of light shines on the top surface of the water, as represented by the ray in the illustration. Draw arrows on the illustration that indicate the beam's path from the top of the water out of the bottom of the container (the walls of the container are made of infinitely thin glass and can be disregarded).

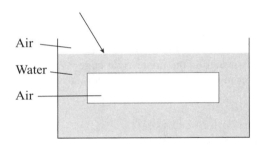

21.2.5 Represent and reason Rays in the illustrations represent beams of light from a laser pointer.

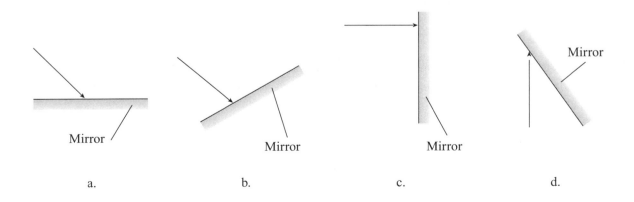

a.   b.   c.   d.

a. Draw on the illustrations reflected rays for each arrangement.

b. Draw on the illustrations two locations for each situation in which you could place a small piece of paper that would be illuminated by the reflected ray.

21.2.6 Explain In Activity 21.1.2 you learned that you only see an object if emitted or reflected light travels from the object to your eye. In Activity 21.1.8 you learned that when a laser beam is reflected off a smooth surface, the incident beam and the reflected beam are at the same angles relative to a line perpendicular to the surface. Imagine that a laser beam hits a wall. If you stand at any place in the room, you see a bright spot on the wall where the laser beam hits it. How can you reconcile these two phenomena? *Hint:* Examine the surface of the wall and compare it to the mirror.

21.3 | Quantitative Concept Building and Testing

21.3.1 Observe and find a pattern Fill a fish tank with water and shine the light from a laser pointer at different angles on the top surface of the water. The angle θ_1 of the incident beam relative to the normal line and the angle of the beam that propagates into the water (the so-called refracted ray θ_2) are shown in the diagram and recorded in the table. Various trigonometric functions of those angles are recorded as well.

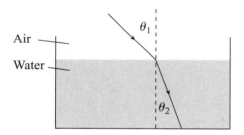

Incident angle θ_1	cos θ_1	sin θ_1	Refracted angle θ_2	cos θ_2	sin θ_2
20°	0.94	0.34	15°	0.97	0.26
30°	0.87	0.5	22°	0.93	0.37
40°	0.77	0.64	29°	0.87	0.48
50°	0.64	0.77	35°	0.82	0.57
60°	0.50	0.87	41°	0.76	0.65

a. Use any or all of the values given in the table to devise a rule that relates the angle of incidence and the angle of refraction. *Hint:* You might see if the ratio of two quantities has the same value for all angles. If so, use this to help devise a rule.

b. While performing the experiment, you see a dot of light on the ceiling. Explain.

21.3.2 Observe and find a pattern
Repeat Activity 21.3.1, only this time imagine that you shine the light through air onto a block of glass with a smooth surface. The table records the angle between the incident light and the normal line and the angle between the refracted light and the normal line.

Incident angle θ_1	Refracted angle θ_2
20°	13°
30°	19°
40°	25°
50°	30°
60°	35°

a. Use the rule relating the angle of incidence and the angle of refraction that you devised in the previous activity, only this time apply it for light propagation from air into glass. Compare and contrast the air–glass refraction with the air–water refraction.

b. Use the wave model of light and the results from part a and Activity 21.3.1 to compare the speeds of light in air, water, and glass. Justify your answer.

21.3.3 Observe and explain
Place a pencil in a glass that is half full of water. Observe the shape of the pencil. Draw a picture of what you observe and indicate where your eye is on that picture. Explain using the pattern that you found in Activity 21.3.1.

21.3.4 Test your idea
We observed earlier in this chapter that light moving from air to glass or to a liquid bends (refracts) toward a normal line that is perpendicular to that surface. Also, light moving in a glass or liquid into air bends away from a line that is perpendicular to that surface. Use this idea to predict qualitatively what happens to a laser beam in each of the experiments below. *Hint:* Do not forget to draw a normal line at the location at which the light beam hits the border of the two media. Use a solid glass prism and a hollow glass prism to complete the table that follows.

Illustration of the experiment	Use your knowledge of refraction to predict qualitatively the path of the beam.	Perform the experiment and record the results (i.e., the path of the beam).	Discuss whether your prediction was successful or if the relationship needs to be modified.
Solid glass prism in air Laser beam → Glass / Air			
Hollow glass prism in water Water Laser beam → Air			
Solid glass prism in water. Note that the light bends toward the perpendicular line going from water to glass, and vice versa in going from glass to water. Water Laser beam → Glass			

21.3.5 Test your ideas Use your knowledge of refraction to predict qualitatively and quantitatively what will happen in the described experiment. Complete the table that follows. Note that the index of refraction of air is 1.0 and of water is 1.33.

Shine a laser beam so that it passes through glass and then refracts into the air above. Vary the angle of incidence on the glass–air interface (e.g., see rays 1 and 2).	Predict what will happen if you gradually increase the angle of incidence. Identify a special angle of incidence where there is no longer a refracted ray. Explain your prediction.	After making the prediction, perform the experiment—did you correctly identify the special angle? What happens at incident angles greater than this special "critical" angle?
Air 2 1 Glass		

21.4 | Quantitative Reasoning

21.4.1 Represent and reason A beam of light hits a plane mirror perpendicular to the mirror's surface. Determine the angle between the incident and reflected beams if you tilt the mirror 30°. Include a sketch of the initial situation before tilting the mirror and the final situation after tilting it. Draw in labeled rays representing the incident and reflected beams for both orientations.

21.4.2 Represent and reason Two mirrors are placed together at a right angle, with one mirror oriented vertically and the other oriented horizontally. A ray strikes the horizontal mirror at an incident angle of 60° relative to the normal line, reflects from it, and then hits the vertical mirror.

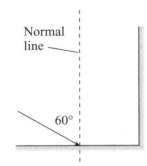

Normal line

60°

a. Determine the angle of incidence relative to the vertical mirror.

b. Use the law of reflection and the drawing to show that the ray leaves the vertical mirror parallel to its original direction.

21.4.3 Represent and reason Imagine that you shine a laser beam at a glass plate, as shown in the illustration. Use your knowledge of reflection and refraction to predict the path of the beam. Draw rays to indicate the path of the beam. (*Hint:* Before you draw the rays, decide on the direction of the normal line at the point at which the laser light first hits the glass.)

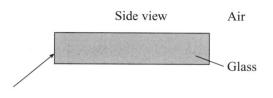

21.4.4 Represent and reason A laser beam shines up through a piece of glass of refractive index 1.56 and reaches the glass–air interface at the top, as shown in the figure.

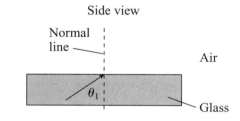

a. At what range of incident angles θ_1 will the laser beam not pass out of the glass and into the air? Explain your prediction.

b. Use your prediction in part a to explain how a glass rod (actually a thin glass fiber) can become a pipe that transmits light signals without light losses out the side walls of the fiber. Indicate any assumptions you made.

21.4.5 Pose a problem You have a block of light crown glass and a laser pointer. The index of refraction of light crown glass is 1.517. Pose a problem for which you need to use the knowledge of refraction and of total internal reflection to solve.

21.4.6 Represent and reason Light enters a right-angle prism as shown in the illustration and experiences total internal reflection. Draw the path of the light on the illustration. What is the minimum refractive index for the prism for total internal reflection to occur?

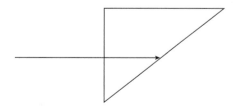

21.4.7 Equation Jeopardy Complete the table that follows.

Mathematical representation	Solve for the unknown(s).	Sketch the situation that the equation describes.	Write a word description of the process.
$1.00 \sin 53° = n_2 \sin 41°$			
$1.00 \sin 53° = 1.56 \sin \theta_2$ and $1.56 \sin \theta_2 = 1.00 \sin \theta_3$			

21.4.8 Evaluate the solution

The problem: The eyes of a person standing at the edge of a 1.2-m-deep swimming pool are 1.6 m above the surface of the water. The person sees a silver dollar at the bottom of the pool at a 37° angle below the horizontal. Determine the horizontal distance d from the person to the dollar.

Proposed solution:

Sketch and translate

The situation is sketched at the right.

Simplify and diagram

The water surface is smooth, and the index of refraction of water is 1.33.

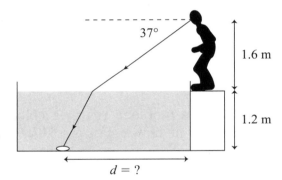

A ray from the eye to the coin is shown in the figure at the right.

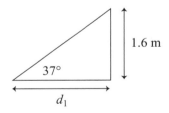

Represent mathematically, solve, and evaluate

The horizontal distance d_1 from the eye to the water surface where the ray enters is determined using trigonometry (see the triangle in the figure at the right).

$$\sin 37° = (1.6 \text{ m})/d_1 \quad \text{or} \quad d_1 = 2.67 \text{ m}$$

Apply Snell's law to find the angle of the ray while in the water:

$$1.00 \sin 37° = 1.33 \sin \theta_{\text{in water}} \quad \text{or} \quad \theta_{\text{in water}} = 26.9°$$

Finally, we can use the triangle in the water to determine the extra horizontal distance d_2 in the water:

$$\sin 26.9° = d_2/(1.2\,\text{m}) \qquad \text{or} \qquad d_2 = 0.54\,\text{m}$$

The total horizontal distance from the edge of the pool under the person's feet is:

$$d = d_1 + d_2 = 2.67\,\text{m} + 0.54\,\text{m} = 3.2\,\text{m}$$

a. Identify any errors in the student solution.

b. Provide a corrected solution if you find errors.

21.4.9 Reason and explain Different colors of light have different indexes of refraction when passing though water droplets (see the table).

Color	Index of refraction
Red	1.613
Yellow	1.621
Green	1.628
Blue	1.636
Violet	1.661

How can this information qualitatively explain why we see a rainbow while it rains? Think of where the Sun is located with respect to an observer when the observer sees a rainbow. What assumption about the shape of the water droplets in the air can we make?

21.4.10 Design an experiment You have a semicircular piece of transparent material. Design two independent experiments to find the index of refraction of this material.

a. Fill in the table that follows.

Experiment 1			
Describe the experiment; draw a ray diagram to depict the experimental setup.	Describe the procedure to determine n.	List experimental uncertainties.	Perform the experiment, record the outcomes, and calculate n.
Experiment 2			
Describe the experiment; draw a ray diagram to depict the experimental setup.	Describe the procedure to determine n.	List experimental uncertainties.	Perform the experiment, record the outcomes, and calculate n.

b. Did the two experiments give you the same value of n? Explain the discrepancies.

21.4.11 Observe and explain Use a glass microscope slide and carefully submerge it in vegetable oil. Complete the table that follows.

Experiment	Describe your observations and suggest an explanation.	Design and perform an experiment to compare the indexes of refraction of oil and glass.	Write an explanation of why we cannot see objects that have the same index of refraction as the medium in which light travels.
As you submerge the slide into oil, the submerged part of the slide disappears.			

22 Geometrical Optics

22.1 | Qualitative Concept Building and Testing

22.1.1 Observe and explain Three friends stand behind a candle that is positioned 20 cm in front of a plane mirror. They observe the image of the candle, and each of them points a ruler in the direction of the image of the tip of the candle they see in the mirror. The dashed lines in the illustration indicate the orientations of their rulers.

Extend the dashed lines behind the mirror to locate the image of the candle relative to the mirror. To the three friends looking at the mirror, light seems to be coming from that image location.	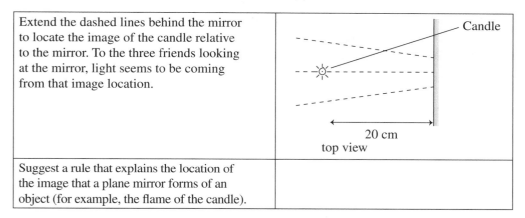
Suggest a rule that explains the location of the image that a plane mirror forms of an object (for example, the flame of the candle).	

22.1.2 Reason and explain Use two arbitrary rays to explain why the image of the candle in Activity 22.1.1 is at the same distance behind the mirror as the candle is in front of the mirror. Complete the table that follows to help you.

Two rays beginning from a point on the candle and moving toward the mirror		The same rays after reflection from the mirror. Do the reflected rays shown above ever meet? If not, how does the mirror form an image?	
Explain why we see an image of the candle behind the mirror.		Extend the reflected rays back behind the mirror to find the image of the candle. Use geometry to prove that the image is the same distance behind the mirror as the candle is in front.	

22.1.3 Test your idea Ari and Samantha are investigating the location of the image of a candle produced by a plane mirror. Ari says that the image is on the surface of the mirror. Test his idea by designing an experiment whose outcome contradicts the prediction based on Ari's idea.

Describe an experiment to test Ari's idea.	Predict the outcome of the experiment based on Ari's idea.	Perform the experiment and record the outcome.	Discuss whether the experiment disproves the idea that the image is on the surface of the mirror.

22.1.4 Observe and explain Shine a light from a laser pointer on a plane mirror and observe the reflected ray. Then shine it on a curved mirror, which is a section of a sphere of radius R (concave). For two examples, the light reflects as shown in the illustration. Explain the behavior in terms of the law of reflection. *Note:* The dashed line R perpendicular to the curved mirror's surface in the illustration passes through the center O of the sphere from which the mirror was cut.

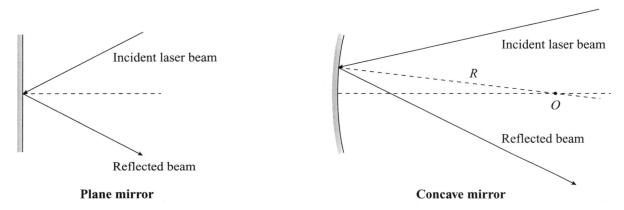

Plane mirror   **Concave mirror**

22.1.5 Observe and explain Shine the light from a laser pointer on a curved mirror, which is a section of a sphere of radius R (convex). The light reflects as shown in the illustration. Explain the behavior in terms of the law of reflection.

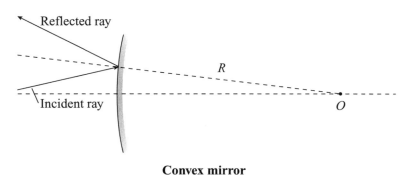

Convex mirror

22.1.6 Observe and explain Assemble a concave mirror and three handheld lasers. Point the beams of the lasers parallel to the main axis of the mirror (a horizontal axis through the center of the mirror). After reflection, the rays all pass through the same point exactly in the middle between the mirror and the center of the sphere from which the mirror was cut. This point is called the *focal point*—the point through which rays parallel to the axis of the concave mirror pass after reflection from the mirror. Use the law of reflection to explain this observation.

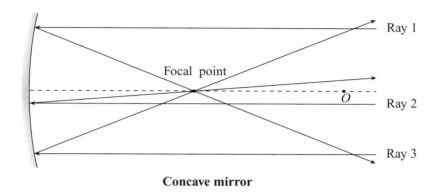

Concave mirror

22.1.7 Test your idea Assemble a concave mirror and two handheld lasers. Fill in the table that follows. *Hint:* Reviewing Activity 22.1.6 might help you make your prediction.

Draw a ray diagram to predict what will happen if you aim the beams of the lasers so that they both pass through the focal point before hitting the mirror.	Perform the experiment and record the outcome. Discuss whether the prediction was confirmed.	Compare and contrast this experiment with the experiment in Activity 22.1.6.

22.1.8 Observe and explain Assemble a concave mirror and three handheld lasers. Aim the laser beams toward the mirror parallel to each other but not parallel to the main axis of the mirror. One of the rays passes through the center of the sphere from which the mirror was cut (point *O*). This ray reflects back along the same path. After reflection, the other two rays pass through the point where the center-passing ray crosses the focal plane of the mirror.

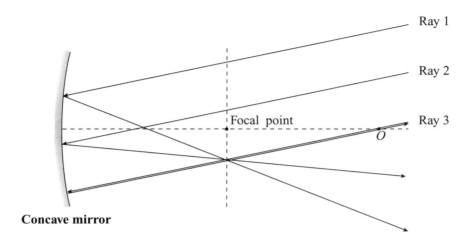

Use the law of reflection to explain this observation.

22.1.9 Test your idea

a. Predict using the law of reflection where the rays parallel to the main axis of a spherical convex mirror of radius *R* will meet after reflection or, alternatively, a point from which all of the reflected rays seem to originate.

b. Use several laser beams to perform the experiment. Record the outcome.

c. Does a convex mirror have a focal point? If so, where?

22.1.10 Observe and explain Shine three parallel beams of light from laser pointers on a thin convex lens made of glass (its surfaces are segments of a sphere). Refracted beams cross at a point on the main axis—called the *focal point*. Explain qualitatively the path of each ray using the law of refraction—what happens at each glass–air interface to cause the net refraction of the rays shown in the illustration?

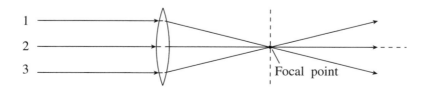

©2014 Pearson Education.

Notice that if you aim the rays from the right side toward the left side of the lens (opposite the way shown in the sketch), after refraction they will pass through another focal point on the left side of the lens.

22.1.11 Observe and explain Shine three parallel beams of laser light onto a thin concave lens made of glass (its surfaces are segments of a sphere, as shown).

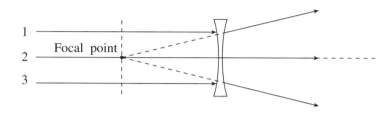

After passing through the lens, the rays diverge and appear to come from a focal point on the main axis. Explain qualitatively the path of each ray using the law of refraction.

22.2 | Conceptual Reasoning

22.2.1 Represent and reason A candle burns in front of a plane mirror, as shown in the illustrations. Consider the flame to be a pointlike source of light. For each case, locate the flame's image by drawing any two rays on the illustration. (Rays can extend in any direction that strikes the mirror.)

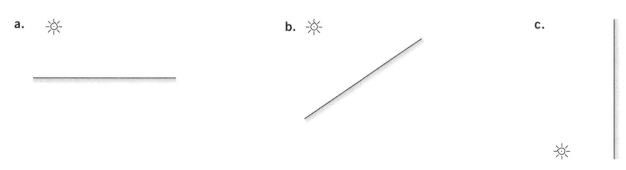

Then devise a rule that can be used to locate plane-mirror images without using rays.

22.2.2 Represent and reason Draw ray diagrams to answer the following questions: How does the size of a plane-mirror image change (increase or decrease) when the object is moved away from a plane mirror? How does the size change when the object is moved toward the plane mirror?

22.2.3 Represent and reason Imagine that you stand in front of a plane mirror to look at your image.

a. Draw a ray diagram to determine the minimum size of a mirror in which you can see your entire body.

b. Where should you put the top of the mirror relative to the top of your head?

22.2.4 Represent and reason Draw a ray diagram showing the path of each of the rays described below after reflection from a concave mirror. Remember that you can locate the image of an object formed by a concave mirror by using any two of these four rays:

a. a ray that is parallel to the main axis of the mirror,	**b.** a ray that passes though the focal point of the mirror,	**c.** a ray that passes through the center of the sphere from which the mirror was cut,	**d.** a ray that is parallel to the ray in part c.

©2014 Pearson Education.

22.2.5 Represent and reason A small, shining object is placed above the main axis of a concave mirror at a distance $s > R$ from the mirror. Two rays are used to find the image of the top of the object. Explain the path of each ray in the illustration.

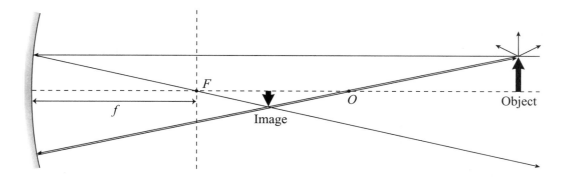

What assumptions were made in the diagram? Point O is the center of curvature of the mirror—that is, the center of the sphere from which the mirror was cut. The focal point is indicated by F, and the focal length is f.

22.2.6 Observe and explain Place a candle or small lightbulb in front of a concave mirror. Use a small white card between the light and the mirror to find a place where a sharp image of the candlelight is formed. Explain the results using a ray diagram.

22.2.7 Represent and reason A ray diagram helps us understand how to find the image of an object produced by a convex mirror. Explain the path of each ray and how we know where the image is located.

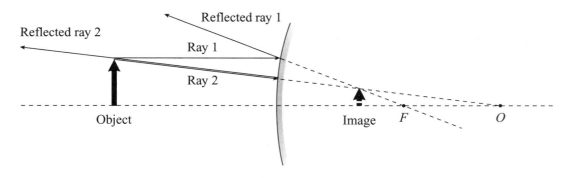

Why is the image drawn above with a dashed line? What assumptions were made in the diagram?

22.2.8 Represent and reason Fill in the table that follows.

Experiment	Sketch	Description
Place a small shining object at a distance s ($s > R$) from a concave mirror. Draw two rays to represent the situation—one parallel to the main axis and one going through the focal point. Draw the reflected rays and find the location of the image of the object.		Describe the image using adjectives: upright (inverted), real (virtual), enlarged (reduced).
Place a small shining object at a distance s ($f < s < R$) from a concave mirror. Draw two rays to represent the situation—one parallel to the main axis and one going through the focal point. Draw the reflected rays and find the location of the image of the object.		Describe the image using adjectives: upright (inverted), real (virtual), enlarged (reduced).
Place a small shining object at a distance s ($s < f$) from a concave mirror. Draw two rays to represent the situation—one parallel to the main axis and one reflected from the center of the mirror. Draw the reflected rays and find the location of the image of the object.		Describe the image using adjectives: upright (inverted), real (virtual), enlarged (reduced).

22.2.9 Represent and reason Use ray diagrams to determine how the size of an image changes (increases or decreases) when the object is moved away from a concave mirror.

22.2.10 Evaluate the reasoning Your friend Brian thinks that if you cover the bottom half of a concave mirror, the real image of a candle produced by the mirror will be cut in half. Do you agree or disagree? If you disagree, how can you convince Brian of your opinion?

22.2.11 Represent and reason Fill in the table that follows.

Place a small shining object at a distance s ($R/2 < s < R$) from a convex mirror. Use a ray diagram to find the location of the image. Describe the image using adjectives: upright/inverted, real/virtual, enlarged/reduced.	Place a small shining object at a distance s ($s < R/2$) from a convex mirror. Use a ray diagram to find the location of the image. Describe the image using adjectives: upright/inverted, real/virtual, enlarged/reduced.

22.2.12 Represent and reason Draw in any two rays to construct images of the objects shown as arrows in the figures below. Confirm by drawing in a third ray. Describe the images using adjectives: upright/inverted, real/virtual, enlarged/reduced.

a.
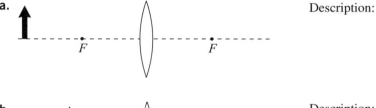
Description:

b.
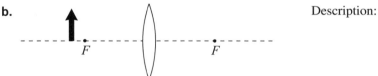
Description:

c.
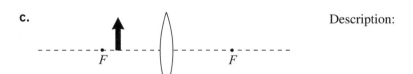
Description:

d.
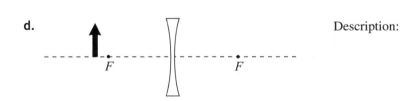
Description:

22.2.13 Evaluate the reasoning Your friend Ritesh says that it's appropriate to call a convex lens a converging lens and a concave lens a diverging lens. How would you convince him that his classification is not always correct?

22.3 | Quantitative Concept Building and Testing

22.3.1 Test the mirror equation Assemble a concave mirror and a candle. Do not light the candle yet.

a. Place the candle behind the focal point away from the mirror. Do not light the candle yet. Fill in the table that follows. *Note:* Use a paper screen between the candle and the mirror to check your prediction.

Measure the distance s between the candle and the mirror. Draw a ray diagram and use the mirror equation to predict the distance s'.	Light the candle, perform the experiment, and measure s'. Record the difference between your prediction and the measured value of s'.

b. Explain the difference between the predicted and measured values of s' (think of experimental uncertainties and assumptions that we made deriving the mirror equation).

22.3.2 Test the lens equation You have a convex lens. Use a source of light that is far away (you can use the sun on a good day or a window in the classroom if it is far enough away—15–20 m) to find a focal length of the lens.

a. Fill in the table that follows.

Draw a ray diagram for the light source and the lens.	Perform the experiment and find the focal length of the lens.

b. Assemble the same lens and a candle. Place the candle behind the focal point and away from the lens, measure the distance between the candle and the lens, and use the lens equation to predict the location of the image of the candle. Use a paper screen to locate the image and hence to check your prediction. Fill in the table that follows.

Measure the distance s.	Predict the distance s'.	Measure the distance s'.	Record the difference between your prediction and the measured values of s'.

c. Explain the difference between predicted and measured values of s' (think of experimental uncertainties and assumptions that we made deriving the lens equation).

22.3.3 Test the lens equation Assemble a convex lens whose focal length is $+10$ cm.

a. Predict where you should hold the lens with respect to a white piece of cardboard placed on your desk to form a focused image of a window onto this piece of cardboard. What mathematical model did you use to make a prediction?

b. Check your prediction and record the result. What is your judgment about the mathematical model that you used to make the prediction?

22.4 | Quantitative Reasoning

22.4.1 Regular problem Use ray diagrams and the mirror equation to locate the position, orientation, and type of image formed by an upright object held in front of a concave mirror of focal length $+20$ cm. The object distances are (a) 200 cm, (b) 40 cm, and (c) 10 cm.

22.4.2 Regular problem A large concave mirror of focal length $+3.0$ m stands 20 m in front of you. Describe the changing appearance of your image as you move from 20 m to 1 m from the mirror. Indicate distances from the mirror where the change in appearance is dramatic.

22.4.3 Represent and reason Using a ruler, carefully draw ray diagrams to locate the images of the objects listed below. Measure the image locations on your diagrams and indicate if they are real or virtual, upright or inverted. When you are done, check your work using the lens equation. (Choose a scale so that your drawing fills a significant portion of the width of the paper.)

a. an object that is 12 cm from a concave lens of -5-cm focal length,

b. an object that is 7 cm from the same lens,

c. an object that is 3 cm from the same lens.

22.4.4 Represent and reason A magnifying glass is a convex lens that when held close to an object (slightly closer than the focal length of the lens) allows you to see its enlarged upright virtual image. Draw a ray diagram to explain how a magnifying glass works.

22.4.5 Represent and reason Imagine you have a $+20$-cm focal-length convex lens. You place an object 15 cm from the lens on the main axis.

a. Fill in the table that follows.

Draw a ray diagram to find the image of the object.	Use the lens equation to calculate the location of the image.

b. Is the calculation consistent with the ray diagram?

c. What is the meaning of the negative sign in the distance of the image?

22.4.6 Represent and reason Imagine that you have a -20-cm focal-length concave lens. You place an object 25 cm from the lens on the main axis.

a. Fill in the table that follows.

Draw a ray diagram to find the image of the object.	Use the lens equation to calculate the location of the image.

b. Is the calculation consistent with the ray diagram?

c. What is the meaning of the negative sign in the distance of the image?

22.4.7 Reason and explain A simple camera system consists of an opening for light, a convex lens, and photographic paper placed where the image of the object is formed.

a. Draw a ray diagram explaining how a camera allows us to produce images of objects.

b. Think of possible technical arrangements of a camera that allow us to make sharp images of objects that are located at different distances from the same lens.

22.4.8 Diagram Jeopardy In the figures below, you see an axis of a lens (the lens itself is not shown) and the location of a shining object and its image. Your task is to find the location and the type of the lens (convex or concave) that could produce this image and find the focal points of the lens. When you think you have found an appropriate lens type and lens location, draw a ray diagram to help justify your choice and show the focal length on the diagram.

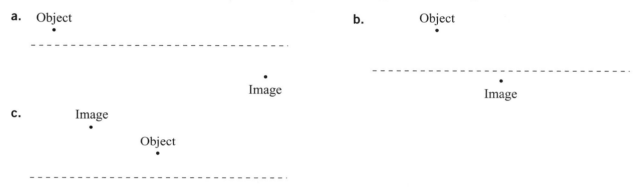

22.4.9 Reason and explain An unlabeled sketch of the optical system of a simple camera is shown below (real cameras have multiple lenses).

The eye is similar in some ways. Complete the table to identify the analogous elements in the eye and the camera.

Purpose	Identify the element for the camera.	Identify the element for the eye.
Allows the light from an object to enter the device		
Allows the light to be focused		
A place where the image is captured		
Makes it possible to get sharp images of objects that are at different distances from the place where the image is formed		
Makes it possible to change the amount of light entering the system		

22.4.10 Represent and reason Sketch an eye with its lens and draw a ray diagram showing how the eye forms an image on the retina. Should the eye lens be convex or concave?

22.4.11 Represent and reason The *far point* of the eye is the greatest distance to an object on which an eye can comfortably focus. The *near point* of an eye is the closest distance of an object on which the eye can comfortably focus. A nearsighted person has trouble focusing on distant objects (such as a sign on the highway). A farsighted person has trouble focusing on objects that are near the eye (such as the morning newspaper). Suggest reasons for these defects of vision and possible ways to correct them. Support your answers with ray diagrams.

22.4.12 Regular problem The image distance for the lens of a person's eye is 2.20 cm. Calculate the focal length of the eye's lens system for an object at the following distances:

a. at infinity,

b. 500 cm from the eye,

c. 25 cm from the eye.

22.4.13 Regular problem A farsighted man can see sharp images of objects that are 3.0 m or more from his eyes. He would like to read a book held 30 cm from his eyes. Determine the focal length of the lenses he needs for his glasses.

He will hold the book 30 cm from his glasses. An eyeglass lens should form an image 3.0 m (300 cm) in front of the lens. The optical system of his eye will look at this image. Draw a sketch of the object (book), eyeglass lens, and the image of that object. Enter known information in your sketch.	
Draw a ray diagram for the eyeglass lens system described in the first cell of the table. It's similar to that of a magnifying glass.	
Use the lens equation to calculate the focal length of the desired eyeglass lens.	

22.4.14 Regular problem A nearsighted woman can focus see sharp images of objects that are 2.0 m or less from her eyes. She would like to read road signs while driving on the turnpike. Determine the focal length of the lenses she needs for her glasses.

She will look at distant signs through her glasses (assume an infinite distance). An eyeglass lens should form an image of this distant object that is 2.0 m (200 cm) in front of the lens. The optical system of her eye will look at this image. Draw a sketch of the distant object, the eyeglass lens, and the image of that object. Enter known information in your sketch.	
Draw a ray diagram for the eyeglass lens system described in the first cell of the table.	
Use the lens equation to calculate the focal length of the desired eyeglass lens.	

22.4.15 Evaluate the solution

The problem: A man who can focus only on objects in the range of 1.6 m to 4.0 m wants to buy a pair of nonprescription glasses to wear while reading and another pair to wear while driving.

a. Determine the focal length of the glasses he should buy for reading.

b. Determine the focal length of glasses he should buy for driving.

Proposed solution for part a:

Sketch and translate

A ray diagram of the situation is shown at the right. The image should be 1.6 m from the lenses when the object (the book) is about 0.4 m from the lenses (a comfortable distance to hold a book from the eye).

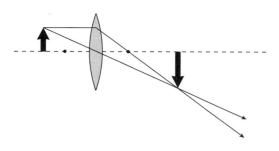

Simplify and diagram

We assumed the book was held 0.40 m from the eye—about 16".

Refer to the ray diagram.

Represent mathematically and solve

Using the lens equation, we can find the focal length.

$$\frac{1}{f_{\text{reading}}} = \frac{1}{0.4} + \frac{1}{1.6} = +2.0 \text{ m}$$

Proposed solution for part b:

Sketch and translate

A ray diagram of the situation is shown at the right. The image should be 4.0 m from the lens when the object (a road sign) is far away—an infinite distance from the lenses.

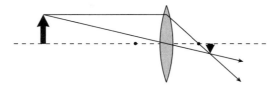

Simplify and diagram

We assume that the object is at infinity.

Refer to the ray diagram.

Represent mathematically and solve

Using the lens equation, we can find the focal length for the glasses used for distance work.

$$\frac{1}{f_{\text{driving}}} = \frac{1}{\infty} + \frac{1}{4.0} = +0.25 \text{ m} = 25 \text{ cm}$$

a. Identify any errors in the proposed solutions to this problem.

b. Provide corrected solutions if you find errors.

23 Wave Optics

23.1 Qualitative Concept Building and Testing

Competing models of light: According to the *particle model of light,* supported by Isaac Newton, some light phenomena can be explained by assuming that light behaves as a stream of tiny discrete particles (corpuscles) that move in straight lines through space, like tiny bullets. According to the *wave model of light,* supported by Christaan Huygens, some light phenomena can be explained by assuming that light behaves as if it is a wave. Each of these two models can explain some light phenomena.

23.1.1 Observe and explain Descriptions of several experiments follow.

a. Complete the table to explain the result of each experiment using the two previously identified models of light.

Experiment	Describe what you observe and sketch the situation.	Explain your observations using the particle model of light.	Explain your observations using the wave model of light.
Light a candle in a dark room and place the candle 2–3 m from the wall. Hold a pencil close to the wall.			
Direct the light from a laser pointer at a plane, horizontal mirror at a 60°-incident angle relative to the normal line.			
Direct the light from a laser pointer at the surface of water in a large, clear container at a 60°-incident angle relative to the normal line.	The refracted beam in the water makes an angle that is less than 60° relative to the normal line.		

b. Which experiment(s) can be explained by both models of light?

23.1.2 Test your ideas Assemble a candle and a piece of cardboard (about 20 cm long and 10 cm wide). Cut two large slitlike holes in the cardboard, as shown in part a of the figure.

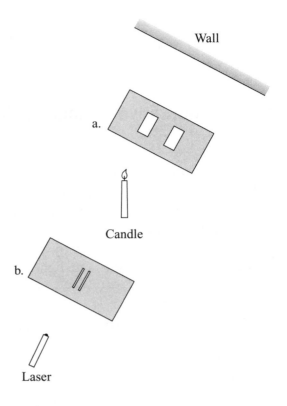

a.

Wall

Candle

b.

Laser

a. Complete the table that follows.

Use the particle model of light to predict what you will see on the wall if you place the cardboard with the large holes between the candle and the wall, situating the cardboard closer to the wall than to the candle.	Use the wave model of light to predict what you will see on the wall if you place the cardboard with the large holes between the candle and the wall, situating the cardboard closer to the wall than to the candle.	Perform the experiment and record the outcome.	Discuss which model gave you a better prediction.

b. Assemble a laser pointer and a plate with two narrow slits situated close to each other. Complete the table that follows.

Use the particle model of light to predict what you will see on the wall if you place the plate with the narrow slits between the laser beam and the wall, situating the plate closer to the laser pointer.	Use the wave model to predict what you will see on the wall if you place the plate with the narrow slits between the laser pointer and the wall, situating the plate closer to the laser pointer. *Note:* Assume that Huygens' wavelets emanate from each slit.	Perform the experiment and record the outcome.	Discuss which model gave you a better prediction.

23.1.3 Explain In Activity 23.1.2b, with the laser pointer and two narrow slits, instead of seeing two thin bright lines (the images of the slits), you saw closely spaced alternating bright and dark narrow bands of light—a bright band at the center with several less-bright bands on the sides. Why were there dark areas between bright areas? *Hint:* Is it possible for two particles to arrive at the same location and cancel each other? Is it possible for two waves to arrive at the same location and cancel each other?

23.2 | Conceptual Reasoning

23.2.1 Reason and explain Waves (any type) are incident from the left side on a barrier with two small openings. Consider what happens on the right side of the barrier. According to Huygens' principle, these openings become wave sources. In the illustration, we represent the wave fronts leaving these sources with dark and light circles. The solid dark lines represent wave crests, and the lighter gray lines represent the troughs beyond the slits at one instant of time.

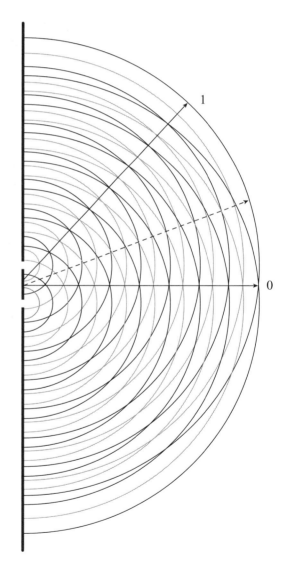

a. Describe what the crests might represent for each of the following types of waves: water waves, sound waves, and light waves.

b. Indicate with the letters "dc" (double crest) places on line 0 that are equal distance from the wave sources (the openings) and where the crests add to form a disturbance that is twice as big as the wave amplitude from one source.

c. Indicate with the letters "dt" (double trough) places on line 0 that are equal distance from the sources and where the wave troughs add to form a negative disturbance whose magnitude is twice the amplitude of a wave trough from one source.

d. The sketch represents the positions of wave crests and troughs at one particular time. Suppose that these are water waves that are now moving and that you stand in the water on the right side of line 0. What would it feel like as the alternating dc and dt points passed you?

e. What if the sketch represents sound waves that are now moving along the line described in part d. What would you hear?

f. What if these were light waves moving along the line described in part d. Would the light be bright or dim? Explain.

g. Would the same effect be observed along line 1 as along line 0? Explain.

h. What would you feel (water waves), hear (sound waves), or see (light waves) if you were located at the end of the dashed line between 0 and 1? Explain.

23.2.2 Reason and explain In Activity 23.1.3, the light and dark bands produced on the wall by laser light passing through two narrow slits are more easily explained using the wave model of light. Consider the wave model and the two-slit phenomenon. Shine the light from a laser pointer onto two closely spaced slits. On a screen several meters to the right of the slits, you observe bright light bands at the positions of the dots shown in the illustration (b_0 at the center, and b_1 and b_2 bands at each side of the center). You see darkness, which we call dark bands, at the positions of the crosses (the d_1 and d_2 bands at each side of the center). *Note:* The separation of the bright and dark bands on the screen is exaggerated in this sketch.

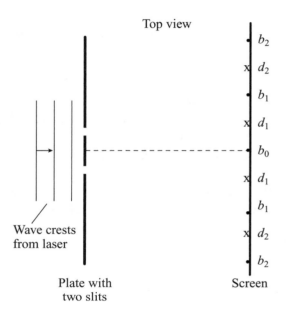

a. Use Huygens' principle to explain why we can assume that the two slits are wave sources and the waves produced by them vibrate synchronously (they are said to be *in phase*).

b. Use the wave model of light to answer the questions in the table that follows. Think about the distances from the two slits to a bright band or a dark band and about superposition of the waves coming from the two slits.

Explain why the center b_0 band is bright.	Explain why b_1 above the center bright band is bright.	Explain why d_1 above the center bright band is dark.
Explain why b_2 above the center bright band is bright.	Explain why d_2 above the center bright band is dark.	

23.2.3 Observe and explain Place a color filter in front of a slide projector so that the beam of light coming out of the projector is of primarily one color. Let the light pass through a single slit whose width can be varied. Place a screen about 1 m beyond the slit. With a 1-cm-wide slit, you see on the screen an image that looks like the slit. Slowly decrease the width of the slit. When the slit width approaches a millimeter (still much wider than the slits in double-slit experiments), a seemingly strange thing happens: The width of the pattern on the screen starts to increase, but the light becomes dimmer. In addition to the widening of the pattern, there are alternating bright and dark fringes. The fringes spread as the slit width decreases. Can you explain these observations with a particle model of light and/or with a wave model of light? *Hint:* You can simplify the situation by using Huygens' principle—that is, by considering the points of the open slit to be tiny light sources emitting half-circle light wavelets in the forward direction. The wavelets can interfere constructively or destructively.

23.3 | Quantitative Concept Building and Testing

23.3.1 Represent and reason Imagine that you shine a laser at a screen with two very narrow, closely spaced slits, as in Activity 23.2.2. These two slits can be considered sources of light wavelets of the same wavelength vibrating in phase—produced by the same wave front arriving at the slits from the left. The illustration shows lines going from each of two slits to the second bright band (b_2) above the central bright band (b_0) on a screen. Dots represent the bright spots, and crosses represent dark spots.

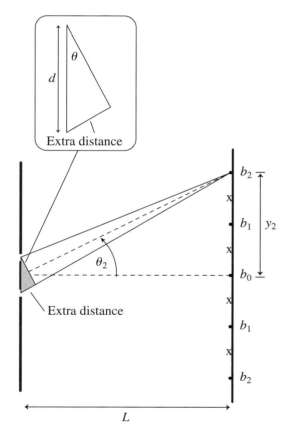

a. Compare the extra distance that light travels from the lower slit to the b_2 bright band and the distance from the upper slit to that bright band. Express this difference in wavelengths of light. Remember that this is the second bright band.

b. How is the angle θ (shown in the inset) related to the angle θ_2 shown in the main part of the figure? Explain.

c. Refer to the triangle insert in the illustration and to the results of Activity 23.2.1 to help you write an expression that relates the extra distance that light travels from the lower slit to the b_2 bright band and the distance from the upper slit to that bright band expressed through the wavelength λ, slit separation d, and the angle θ in the triangle.

d. Write another expression that relates the angle θ_2 to the distance L from the slits to the screen and the distance y_2 from the b_0 central maximum to the position of the b_2 bright spot.

e. Generalize the two expressions developed in parts c and d so that they can be used to determine the angular deflection to the nth bright band. List the assumptions you made when constructing these two expressions.

23.3.2 Test your ideas Gather a laser pointer, a set of double slits of known separations, a screen, and a ruler. Design an experiment to test the relationships you devised in Activity 23.3.1.

a. Complete the table that follows. *Note:* Look at the laser case; it may tell you the wavelength of the light.

Sketch the experimental setup.	Use the expressions devised in Activity 23.3.1c and d to predict the outcome of the experiment.	List additional assumptions you made.	Perform the experiment, record the outcome, and compare it to the prediction.

b. Discuss whether the outcomes of the experiments support the expressions you devised in 23.3.1c and d.

23.3.3 Observe and explain The slit separation for double slits typically used in lecture demonstrations is about 0.5 mm $= 0.5 \times 10^{-3}$ m. An apparatus called a *grating* has about 200 slits in 1 mm.

a. Determine the distance between the centers of the adjacent slits in such a grating.

b. Place the grating about 2 m from a white screen. Shine laser light through the grating and observe a set of bright dots on the screen. These dots are much farther apart than when you use a double-slit apparatus. Draw a diagram of the experimental situation and explain the phenomenon qualitatively.

23.3.4 Reason and explain In Activity 23.3.3, you found a large angular deflection to the first bright band. Suppose you have a five-slit grating with adjacent slits having a small slit separation. The position of the first b_1 bright band to the side of the central b_0 bright band for laser light passing through this grating is shown in the illustration. In terms of light wavelength, how do the distances from second, third, fourth, and fifth slits to the first b_1 bright band compare to the distance from the first (bottom) slit to that bright band? Explain, and be specific.

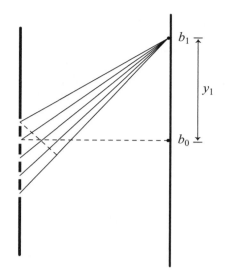

23.3.5 Derive Consider the situation depicted in Activity 23.3.4.

a. Devise a mathematical expression that relates the angular deflection θ_1 to the first bright band, the wavelength λ of the light, and the separation d of adjacent slits. You might want to review what you did in Activity 23.3.1.

b. Devise a mathematical expression that relates the angular deflection θ_1 to the first bright band on the screen to the distance L of the grating from the screen and the distance y_1 of the first bright band from the central maximum b_0 on the screen.

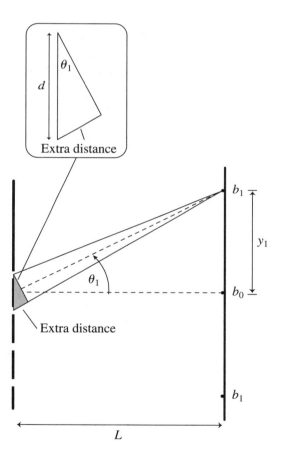

23.3.6 Test your idea

a. Use the expressions that you devised in Activities 23.3.4 and 23.3.5 to predict how an increase in the number of slits per millimeter in the grating (for example, 400 slits per mm instead of 200) should affect the separation of the bright dots on the screen, assuming you use the same laser as in the previous activities. Perform the experiment and compare the outcome to your prediction.

b. Use the explanations that you devised in Activities 23.3.4 and 23.3.5 to predict how an increase in the total number of slits without changing the number of slits per unit length should affect the separation of the bright dots on the screen, assuming you use the same laser as in the previous activities. Create an experiment that you can perform to check your prediction. Then perform the experiment and record the outcome. Was it consistent with your prediction?

23.3.7 Observe and find a pattern
Repeat the experiment in Activity 23.3.3, but this time instead of a laser, use a flashlight and a grating with the greatest number of slits/mm that is available. Complete the table that follows.

Sketch the experimental setup.	Use colored pencils to draw the pattern you observe on the screen.	Describe the pattern in terms of color location.	Write an explanation for the pattern.

23.3.8 Design an experiment
Use the apparatus from Activity 23.3.7 to design an experiment to determine the wavelengths of different colored light coming from an incandescent lightbulb.

Sketch the experimental setup.	List the quantities you will measure.	Summarize the mathematical procedure you will use to calculate wavelengths. Then calculate the wavelengths.	List representative calculated wavelengths of the different colored light.

23.3.9 Derive In previous activities involving two or more slits, we used very narrow slits and considered them pointlike wave sources. What happens if an obstacle with a single slit blocks the light? Imagine that a wave approaches a single slit from the left (see the illustration). The opening is about the same width as the wavelength of the wave. We observe on a screen to the right of the slit a bright band of light in the center (b_0) and alternating dark and bright bands on each side of the center bright band (d_1 and b_1 in the sketch—there are usually more than one dark and bright band on the sides).

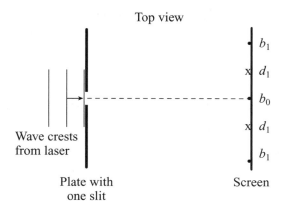

Top view

Wave crests from laser

Plate with one slit

Screen

a. To understand the location of the first dark band (d_1) on the screen, divide the slit in half. What condition is necessary for a wavelet from a point at the top of the top half of the slit to interfere *destructively* at d_1 on the screen with a wavelet from the top of the bottom half of the slit opening? Explain. (The sketch should help.)

b. Write an expression that relates the slit width w, the wavelength of the light λ, and the angle θ_1. Refer to the illustration.

c. If you consider a wavelet produced a little lower in the top half of the opening and another a little lower in the bottom half of the opening, will they also interfere destructively? Explain. In fact if the condition in part a is satisfied, does a wavelet from each point in the bottom half of the single slit interfere destructively with a wavelet from each point in the top half of the slit? Explain.

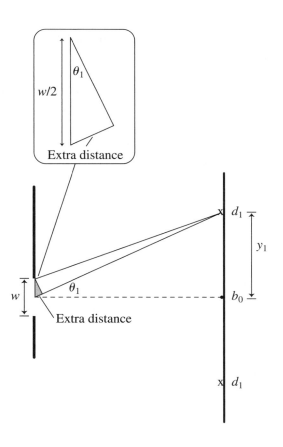

θ_1

$w/2$

Extra distance

w

θ_1

Extra distance

d_1

y_1

b_0

d_1

23.3.10 Test your idea Aim a laser pointer toward a single narrow slit that is part of a system of slits with different widths. Use the relationship derived in Activity 23.3.9 to predict the width of the central maximum on a screen placed about 1 m away from the slit. Complete the table that follows.

Sketch the pattern on the screen for the narrowest slit.	Sketch the pattern on the screen for a medium-width slit.	Sketch the pattern on the screen for the widest slit.
Calculate the distance between the dark bands on each side of the central maximum for each slit.	**Perform the experiment and compare the results with predicted values.**	

23.3.11 Test an idea Babinet's principle states that the diffraction pattern of complementary objects is the same. For example, a slit in a screen produces the same diffraction pattern as a screen the same size as the slit; a hair will produce the same diffraction pattern as a slit of the same width as the hair. Use Babinet's principle to predict the difference in the patterns on a screen produced by laser light shining on a strand of your hair and then shining on a thin sewing needle. Explain your prediction. Perform the experiment and record the outcome. Did the prediction match the outcome of the experiment?

23.4 | Quantitative Reasoning

23.4.1 Regular problem Light of wavelength 540 nm from a green laser is incident on two slits that are separated by 0.50 mm; the light reaches a square screen 50 cm × 50 cm that is 1 m away from the slits. Describe quantitatively the pattern that you will see on the screen. Complete the table that follows.

Sketch and translate	Simplify and diagram
• Visualize the situation and sketch it. • Translate givens into physical quantities.	• Decide if the small-angle approximation is appropriate. • Represent the situation with a ray diagram showing the path of light waves from the two slits to the screen.
Represent mathematically	Solve and evaluate
• Describe the diagram mathematically.	• Solve the problem and decide if the answer makes sense.

23.4.2 Represent and reason Monochromatic light passes through two slits and then strikes a screen. The distance on the screen between the central maximum and the first bright fringe at the side is 2.0 cm.

a. Sketch the situation. Include symbols of quantities used in the equations.

b. Determine the fringe separation if the slit separation is doubled and everything else remains unchanged.

c. Determine the fringe separation if the wavelength is doubled and everything else remains unchanged.

d. Determine the fringe separation if the screen distance is doubled and everything else remains unchanged.

23.4.3 Evaluate the solution

The problem: Determine the width of a hair that when irradiated with laser light of wavelength 630 nm produces a diffraction pattern on a screen with the first minimum 2.5 cm from the central maximum. The screen is 2.0 m from the hair.

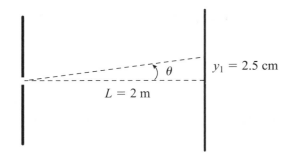

Proposed solution: A hair is a very thin obstacle in the path of light; the light bends around it and produces a diffraction pattern on the screen. According to Babinet's principle, the pattern will be similar to that formed by light passing through a single narrow slit whose width is the same as the width of the hair. Thus, we can use the expression for the angular deflection to the first minimum to relate the angular width of the central maximum to the width of the hair. Because the angular deflection is small, the sine and tangent of this angle give the same result:

$$\sin\theta_1 = \tan\theta_1 = y_1/L$$

$$w\sin\theta_1 = n\lambda \quad \text{for} \quad n = 1$$

$$w = \frac{\lambda}{\sin\theta_1} = \frac{\lambda y_1}{L} = \frac{(630 \times 10^{-9})(2)}{(2.5)} = 504 \times 10^{-9}\,\text{m}$$

a. Identify any missing elements or errors in the solution.

b. Provide a corrected solution if there are errors.

23.4.4 Regular problem A reflection grating reflects light from adjacent lines in the grating instead of allowing the light to pass through slits, as is the case with the so-called transmission gratings we have been studying. Interference between the reflected light waves produces *reflection maxima*. The angular deflection of bright bands, assuming perpendicular incidence, is calculated using the same equation as the angular deflection of transmitted light through a regular grating. White light is incident on the wing of a Morpho butterfly (whose wings act as a reflection grating).

a. Explain why you see different color bands coming from the wings of the butterfly when white light shines on the wings.

b. Red light of wavelength 660 nm is deflected in first order (n = 1) at an angle of 1.2°. Determine the angular deflection in first order (n = 1) of blue light (460 nm).

c. Determine the angular deflection in third order (n = 1) of yellow light (560 nm).

23.4.5 Design an experiment You have probably noticed that stars have different colors—some are white, some are yellow, and some are red. Does this mean that stars of red color do not emit any blue light? Astronomers use an instrument called a *spectrograph* to analyze the color composition of starlight. The central mechanism of a spectrograph is a grating. Design a simple version of a spectrograph, an apparatus that will allow you to separate different colors of light emitted by a lamp on your desk or from a distant star and will also allow you to measure the wavelengths of these different colors. Draw a picture of the apparatus and explain how it works.

23.4.6 Design an experiment Design an experiment to determine the thickness of an individual strand of your hair using a laser pointer, a screen, and a ruler.

a. Describe the design, the procedure, and the assumptions that you make.

b. Perform the experiment, record the measured quantities, and calculate the width of a piece of hair. Then measure the hair strand with a caliper and decide whether the result you obtained with the first method agrees with the second.

23.4.7 Estimate Assume that Earth, its structures, and its inhabitants are all decreased in size by the same factor. *Estimate* the decrease required so that the first-order diffraction dark band of 500-nm light entering a typical room window is at 90° (the central bright band would light most of the room). Explain all aspects of your calculations and the assumptions that you made.

23.4.8 Design an experiment Examine the apparatus in the illustration to answer the questions that follow.

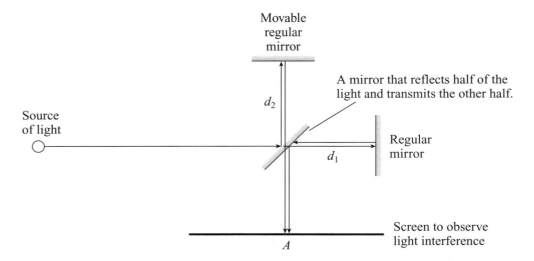

a. Explain why you see an alternating bright light and then a dark spot on the screen as the movable mirror is moved.

b. Explain how this apparatus can be used to determine if the speed of light in one direction is different from the speed of light in another direction.

24 Electromagnetic Waves

24.1 | Qualitative Concept Building and Testing

24.1.1 Observe and find a pattern Several experiments involving traveling waves on a Slinky are described below.

a. Construct a sketch for each experiment. The outcome is shown in the right column.

Experiment	Sketch	The outcome
A Slinky rests on its side on a table. One end is held off the edge of the table. The Slinky is shaken left to right, parallel to the table. Make a sketch of the process.		The pulse continues across the table.
The same as above except the Slinky is shaken up and down, perpendicular to the table.		The pulse does not continue across the table.
The same as above except the Slinky is shaken forward and backward, along the direction of the Slinky. Make a sketch of the process.		The pulse continues across the table.
The same as above except part of the Slinky on the table passes through a tube. The Slinky is shaken forward and backward, a longitudinal pulse. Make a sketch of the process.		The pulse continues across the table.

b. Devise a rule or rules that describe the pattern of conditions under which the pulse can continue across the table.

24.1.2 Test an idea Light is either a transverse wave or a longitudinal wave. Several experiments that test these two hypotheses are described below. The experiments make use of a polarizer, a device made from material that prevents light waves from passing through if something in the wave is vibrating perpendicular to the axis of the polarizer.

a. Make predictions based on each of the two hypotheses about the brightness of the light once it has passed through the polarizer(s).

Experiment	Prediction if light is transverse wave	Prediction if light is longitudinal wave	Outcome
Light from a light bulb shines on a polarizer and its brightness is detected on the other side.			The light reaching the other side of the polarizer is significantly dimmer.
The same as above except the polarizer is slowly rotated.			The light reaching the other side of the polarizer is significantly dimmer and does not change as the polarizer is rotated.

Experiment	Prediction if light is transverse wave	Prediction if light is longitudinal wave	Outcome
Light from a light bulb shines on a polarizer. A second polarizer is positioned behind the first one. The second polarizer is slowly rotated relative to the first.			The light is dimmer overall but also fades in and out completely as the second polarizer is rotated.

b. Make a judgment about each of the two hypotheses. Which (if any) of them are disproved by these experiments?

24.2 | Conceptual Reasoning

24.2.1 Summarize You learned in a previous chapter that a changing magnetic field produces an electric field. Describe experimental evidence for this.

24.2.2 Represent and reason Antennas are used to produce electromagnetic (EM) waves. This is accomplished by connecting the antenna to a source of an alternating emf. In this activity you will represent the production of an EM wave using field line diagrams. The figures below show the current and charge separation in the antenna at various clock readings.

a. Add appropriate electric and magnetic field lines to these figures. Be sure to include an electric and magnetic field line that passes through the indicated point P.

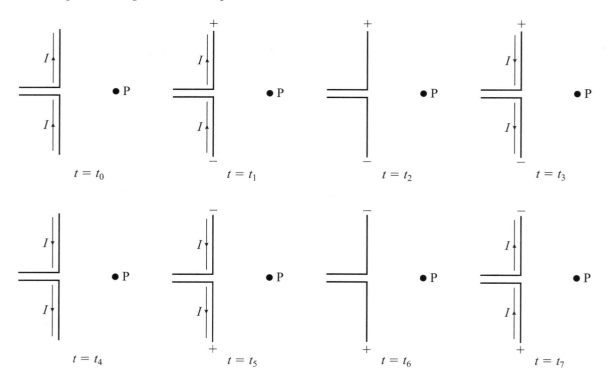

b. For each clock reading t_0 through t_7 draw a vector representing the $\vec{E}$ field at the point P.

• • • • • • • •
t_0 t_1 t_2 t_3 t_4 t_5 t_6 t_7

c. For each clock reading t_0 through t_7 draw a vector representing the $\vec{B}$ field at the point P.

• • • • • • • •
t_0 t_1 t_2 t_3 t_4 t_5 t_6 t_7

d. Describe the behavior of the $\vec{E}$ field and $\vec{B}$ field at the point P.

24.3 | Quantitative Concept Building and Testing

24.3.1 Observe and find a pattern
Experiments involving light waves passing through one or more polarizers are described in the table below. The light intensity is measured using a device known as a photometer.

Experiment	Outcome of experiment
a. Light from a light bulb shines on a polarizer. The polarizer is slowly rotated. The light intensity on each side of the polarizer is measured.	
b. Light from a light bulb shines on a polarizer. A second polarizer is positioned behind the first one. The second polarizer is slowly rotated relative to the first. The light intensity on each side of the polarizer set is measured.	

Experiment	Outcome of experiment
c. Light that has already passed through a vertically oriented polarizer shines on a second polarizer. The second polarizer is slowly rotated. The light intensity on each side of the polarizer is measured.	

For each of the above cases, devise a mathematical relationship between the light intensity I_0 before the polarizer(s) and the light intensity I after.

a.

b.

c.

24.3.2 Derive
The constant k (Coulomb's constant) and the constant μ_0 (vacuum permeability) have appeared in your study of electric and magnetic phenomenon.

a. By analyzing the units of these two constants, devise a mathematical expression involving these two constants that has units of speed, m/s, and determine its value.

b. What is the significance of this value?

24.3.3 Represent and reason
An electromagnetic wave traveling in the positive x direction can be represented by the following wave equations for the $\vec{E}$ field and $\vec{B}$ field.

$$E_y = E_{max} \cos\left[2\pi\left(\frac{t}{T} - \frac{x}{\lambda}\right)\right] \quad B_z = B_{max} \cos\left[2\pi\left(\frac{t}{T} - \frac{x}{\lambda}\right)\right]$$

To see if these equations are reasonable, draw graphs of E_y as a function of x at the following clock readings: $t = 0, t = T/4, t = T/2$, and $t = 3T/4$. Assume the wave is a signal from a 2100 MHz 3G cell phone tower.

Graph E_y as a function of x at the given clock readings	
$t = 0$	$t = T/4$
$t = T/2$	$t = 3T/4$

24.3.4 Determine everything you can Below is the graph of the $\vec{B}$ field component of an electromagnetic wave. Write down everything that you can about this wave.

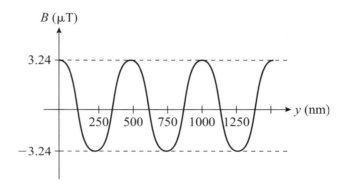

24.3.5 Evaluate The expressions for the electric and magnetic field energy densities are

$$u_E = \frac{1}{2}\varepsilon_0 E^2 \ \text{ and } \ u_B = \frac{1}{2\mu_0}B^2$$

Check that the units of these expressions are consistent.

24.3.6 Observe and find a pattern You are working late on a cold winter night. The light from a rising full Moon reflects off a nearby frozen fountain and reaches your eyes. The glare is very bright so you put your sunglasses on. As you do this you notice that as you rotate your sunglasses the reflected light of the Moon passing through the sunglasses varies in brightness. You head outside to investigate this and record your observations shown in the table that follows.

Experiment	Outcome of experiment	
As the Moon rises (decreasing the angle of incidence) you observe the light passing through the rotating sunglasses	θ_i, θ_{refl} (°)	Brightness variation
	35	Moderate
	45	Large
	53	Maximal (brightness goes to zero for some orientations)
	65	Large

a. Analysis of observations.

θ_{refr} (°), from Snell's law	Ray diagram (use a protractor for the angles for accuracy)

b. Devise a rule or rules that describe the conditions under which the reflected light is completely polarized in the plane parallel to the reflecting surface. *Hint: Look at each ray diagram and the corresponding values of the angles of reflection and refraction.*

24.3.7 Derive The pattern discovered in the preceding activity can be expressed more conveniently as an equation for the so-called polarization angle θ_p, the angle of incidence at which the reflected light is completely polarized in the plane of the reflecting surface. Follow these steps to derive this relationship:

1. Begin with Snell's law.
2. Use the third ray diagram above under Activity 24.3.6 (a) to come up with an equation involving angles in the diagram that relates θ_2 and θ_p.
3. Use this equation to substitute for θ_2 in Snell's law.
4. Finally, use the trigonometric identity $\cos\theta = \sin(90° - \theta)$ and simplify the result as much as possible to arrive at what is known as Brewster's law.

24.3.8 Test an idea See the preceding two activities. The next day you are again working late. You see that a piece of Plexiglas® sheet has been placed over the fountain. This is an opportunity to test the idea developed in the preceding activities.

a. Analyze the following experiment:

Experiment	Prediction of reflection angle for completely polarized reflected moonlight off the Plexiglas sheet using equation from Activity 24.3.7.	Outcome of experiment
The light from a rising full moon reflects off the Plexiglas sheet and reaches your eyes.		54° ± 3°

b. Make a judgment about the idea being tested. Has it been disproved by this experiment?

24.4 ■ Quantitative Reasoning

24.4.1 Estimate In 1899 the first transmission of information across the English Channel via radio waves was achieved. Estimate the time interval between when the signal was produced and when it was received.

24.4.2 Equation Jeopardy Come up with a problem that is consistent with the solution below.

$$\frac{I}{I_0} = \frac{1}{4} = \cos^2(\theta)$$

$$\frac{1}{2} = \cos(\theta)$$

$$\theta = \cos^{-1}\left(\frac{1}{2}\right) = 60°$$

24.4.3 Analyze A secret mission needs to be flown by the military into a neighboring country. There are two radar stations that the aircraft must fly between. They are 60 km apart. The radio waves emitted by station 1 have a period of 2.0×10^{-4} s and a pulse width of 12.0 μs. The waves emitted by station 2 have a period of 2.7×10^{-4} s and a pulse width of 10.0 μs. Is it possible for the aircraft to fly undetected into the neighboring country? Justify your answer quantitatively.

24.4.4 Analyze Some mobile phone carriers offer 4G networks in major metropolitan areas. One version of 4G uses a carrier wave frequency of 700 MHz.

a. How does this frequency compare with the frequencies of visible light?

b. What is the wavelength of this carrier wave?

c. How does this wavelength compare with the wavelengths of visible light?

24.4.5 Evaluate the solution Determine the magnetic field between the plates of a parallel plate capacitor. The capacitor is connected to a 12-V source of EMF, which keeps the plates at a constant potential difference of about 12 V. The plates of the capacitor are separated by 1.0 cm. State any assumptions that you make.

Solution:

$$E = \frac{\Delta V}{d} = \frac{12 \text{ V}}{1.0 \text{ cm}} = 12 \text{ N/C}$$

$$B = \frac{E}{c} = \frac{12 \text{ N/C}}{3.0 \times 10^8 \text{ m/s}} = 4.0 \times 10^{-8} \text{ T}$$

This assumes the current in the circuit is constant.

a. Identify any errors in the solution.

b. Provide a corrected solution if there are errors.

24.4.6 Determine everything you can A satellite broadcasting a satellite television signal orbits at an altitude of 22,300 miles above Earth's surface. It broadcasts radio waves with a total power output of 200 W per channel. Receivers designed to tune in to this signal are dish-like in shape and have a radius of about 40 cm. Determine as many physics quantities relevant to this situation as you can.

25 Special Relativity

■

25.1 | Qualitative Concept Building and Testing

25.1.1 Observe and find a pattern In the following experiments the speed of physical objects and of light when bouncing off objects is investigated.

Experiment	Sketch	Outcome of experiment
A cart moving to the right at 2 m/s on an air track bounces off a barrier at the end.	Cart Barrier $\vec{v_i}$ $\vec{v_f}$	The cart now travels to the left at 2 m/s.
As above but the barrier now moves to the left at 1 m/s relative to the track.	Cart Moving barrier $\vec{v_i}$ $\vec{v_f}$	The cart travels to the left at 4 m/s after bouncing (the cart was traveling at a speed of 3 m/s relative to the barrier both before and after the collision).
A laser is fired toward a mirror 10 m away.	Laser Mirror ── 10 m ──	The laser light returns 6.673283×10^{-8} s later.
As above, but the mirror is being spun so that it is moving toward the laser at 100 m/s.	Laser Moving mirror ── 10 m ──	The laser light returns 6.673283×10^{-8} s later.

Analysis Draw a conclusion about the speed of light. *Hint:* Make an analogy between the cart and the laser light. Also, remember that the index of refraction of air is 1.0003 which has an effect on the speed of light.

25.1.2 Test an idea The experiment below tests the idea that the speed of light is independent of reference frame (it has a value of 299,792,458 m/s in vacuum). The experiment involves an unstable elementary particle that emits electromagnetic waves when it decays.

a. Analyze the experiment and make predictions using the idea under test.

Experiment	Prediction of EM wave travel time to the detector (based on speed of light being independent of reference frame)	Prediction of EM wave travel time to the detector (based on speed of light being relative to the emitting object)	Outcome of experiment
A beam of unstable particles moves to the right at 10^8 m/s. The particles decay 5 m to the right of an EM wave detector.			The average travel time of the EM waves is $(1.7 \pm 0.4) \times 10^{-8}$ s

b. Based on the results of this testing experiment, what is your judgment about these two hypotheses? Explain your reasoning.

25.2 | Conceptual Reasoning

25.2.1 Reason The table below describes various phenomena being observed from several different reference frames. For each phenomenon, determine which reference frame(s) is the proper reference frame. Then, for each of the other reference frames determine whether the time interval between the events will be longer, shorter, or the same as the proper time interval.

Phenomena	Events	Reference frames		
1. A cheetah runs across the savannah, its heart beating.	Successive beats of the cheetah's heart.	**a.** A tourist standing on the ground watching.	**b.** The cheetah.	**c.** A hippopotamus floating in the water where the cheetah stops for a drink.
2. Two people are playing catch. One person throws the ball at a 45 degree angle. The ball is caught by the other person.	The ball being thrown, and the ball being caught.	**a.** The ball.	**b.** The person throwing the ball.	**c.** A person standing on the spot below where the ball reaches its greatest height.
3. You drive from San Francisco to Los Angeles, using your GPS to find your way.	You leave San Francisco; you arrive at Los Angeles	**a.** The GPS satellites.	**b.** You.	**c.** The road you are driving on.

Answers:

Phenomenon	Proper reference frame	Time interval from other reference frames (specify which frame)	
(1)			
(2)			
(3)			

25.2.2 Reason Aaron is able to throw a baseball at 80 mph. If Aaron were to do this from the back of a pickup truck moving at 50 mph with respect to the ground, then the baseball would be moving at 130 mph with respect to the ground. This seems very reasonable. But, what if the truck were moving at $0.5c$ and Aaron could throw the baseball at $0.6c$? Going further, what if Aaron were shooting a laser pointer rather than throwing a baseball? Discuss the relevant issues here and suggest ideas that might resolve them.

25.2.3 Reason A deep space probe is positioned halfway between two stars, A and B, each 1 light-year away. These stars are near the end of their life cycle and are each due to become supernovae soon. A scientist in a spacecraft is traveling from star A to star B. Just as the scientist passes the probe he sees both stars turn supernova at precisely the same time! "What a coincidence! Each star must have exploded precisely one year ago." He also decides to download the probe's data. To his astonishment, the probe reports that the two stars did not explode simultaneously. Use your understanding of relativity to carefully explain how this is possible.

25.2.4 Reason When you studied magnetism you learned that two parallel wires with electric currents in the same direction would exert an attractive magnetic force on each other. One current produces a magnetic field, which in turn exerts a magnetic force on the other current. This explanation is being made from a reference frame at rest with respect to the wires. What about explaining it from a reference frame at rest with respect to the current? Because the charged objects that make up the current are at rest in this reference frame, all magnetic forces exerted on them will be zero!

To resolve this, choose the system of interest to be a single free electron in one of the wires. Assume that the other wire is made of both positively and negatively charged objects. Consider (1) what the motion of these objects is relative to the system, and (2) the phenomenon of length contraction.

25.3 | Quantitative Concept Building and Testing

25.3.1 Derive Muons produced in the upper atmosphere by cosmic rays reach Earth's surface because their lifetimes are extended in Earth's reference frame by time dilation. But, how can this be explained in a muon's reference frame where the muon's lifetime is not extended? Is it possible that the distance the muon must travel to reach Earth's surface is *less* in the muon's reference frame? Let's examine the situation.

a. First, write a kinematics equation describing the motion of the muon in Earth's reference frame. Assume the muon moves in a straight line with constant velocity.

b. Next, write a corresponding equation describing the motion of Earth's surface in the muon's reference frame.

c. The velocities in each of these equations have the same magnitude. Use this to combine the equations.

d. The two time intervals in the equation are related through time dilation. Use this to arrive at a new equation relating the distance the muon travels in Earth's reference frame to the distance Earth's surface travels in the muon's reference frame.

You've just used the ideas of relativity to learn that distances depend on reference frame, a phenomenon known as length contraction.

25.3.2 Analyze The one-dimensional classical velocity transformation equation is:

$$v_{OS,x} = v_{OS',x} + v_{S'S,x}$$

The one-dimensional relativistic velocity transformation equation is:

$$v_{OS,x} = \frac{v_{OS',x} + v_{S'S,x}}{1 + \dfrac{v_{OS',x}v_{S'S,x}}{c^2}}$$

The goal of this task is to determine when it is reasonable to use the simpler classical equation, and when the relativistic equation should be used.

a. First, devise a situation where velocity transformation is relevant. Describe this situation.

b. Now, use a limiting case analysis to investigate the circumstances under which the classical and relativistic equations produce significantly different results (a 5% or greater difference, just to be specific).

25.3.3 Design an experiment Your goal is to design an experiment that will test the following hypothesis:

Doubling the momentum of an object doubles its velocity.

a. Describe your experimental design. Include the prediction and any assumptions made.

b. Explain why your experiment is likely to succeed.

25.3.4 Represent and reason In this activity, momentum is considered as a function of the speed of an object.

a. On a single graph, plot the magnitude of the momentum of an object as a function of its speed 1) using the classical equation and 2) using the relativistic equation. The range of speeds covered by the graph should go *beyond* the speed of light.

b. Discuss consistencies/inconsistencies between the two functions. Discuss any interesting features they have.

c. Repeat part (a), but for the kinetic energy of an object.

d. Repeat part (b) but for the kinetic energy of an object.

25.4 | Quantitative Reasoning

25.4.1 Estimate During his working career, David commutes 1 hour to work and 1 hour back. Anya, however, works from home. Each is given a very precise watch that they carry with them throughout their careers. After they retire, will Anya find David's watch to be behind hers, ahead of hers, or synchronized with hers? Explain. If David's watch is not synchronized with hers, estimate the difference between the clock readings of the two watches.

25.4.2 Equation Jeopardy Come up with a problem that is consistent with the solution below.

$$24 \text{ hr} = \frac{\Delta t_0}{\sqrt{1 - ((2.8 \times 10^8 \text{m/s})/c)^2}}$$
$$\Rightarrow \Delta t_0 = (24 \text{ hr})\sqrt{1 - ((2.8 \times 10^8 \text{m/s})/c)^2}$$
$$\Rightarrow \Delta t_0 = 8.6 \text{ hr}$$

25.4.3 Equation Jeopardy Come up with a problem that is consistent with the solution below.

$$2(2.4 \times 10^{-28}\,\text{kg})c^2 = \frac{2(9.1 \times 10^{-31}\,\text{kg})c^2}{\sqrt{1 - (v/c)^2}}$$

$$\Rightarrow \sqrt{1 - (v/c)^2} = \frac{9.1 \times 10^{-31}\,\text{kg}}{2.4 \times 10^{-28}\,\text{kg}}$$

$$\Rightarrow (v/c)^2 = 1 - \left(\frac{9.1 \times 10^{-31}\,\text{kg}}{2.4 \times 10^{-28}\,\text{kg}}\right)^2$$

$$\Rightarrow v = 0.999993c$$

25.4.4 Analyze At what speed must a proton be traveling so that its kinetic energy is equal to its rest energy? What about an electron? Explain the relationship between your two answers.

25.4.5 Regular problem A futuristic starship is going to be used to travel to Alpha Centauri (4 light-years away from Earth). The starship has two stages. The first stage accelerates the ship to a speed of $0.5c$ relative to Earth. The first stage then separates from the rest of the ship and the second stage activates. The second stage accelerates the ship to a speed of $0.9c$ relative to the first stage.

a. Determine how much time passes on Earth during the journey, assuming the acceleration phases of the ship take only a small fraction of the total travel time.

b. How much time passes on the ship?

c. What distance does the ship travel in its reference frame?

25.4.6 Evaluate the solution A starship moves away from Earth at 40% of light speed. After traveling for 1 month (Earth time) the starship launches a probe back towards Earth at 50% of light speed with respect to the ship. How fast is the probe traveling with respect to Earth?

Solution:

$$v_{PE} = v_{ES} + v_{PS} = 0.5c + 0.4c = 0.9c$$

a. Identify any errors in the solution.

b. Provide a corrected solution if there are errors.

25.4.7 Analyze A galaxy that emits primarily blue light appears primarily red according to detectors here on Earth. Explain how this is possible and describe in as much quantitative detail as you can the motion of the galaxy that would result in this happening.

25.4.8 Regular problem The sun has a total power output of about 4×10^{26} W. How much does its mass decrease each year? What percentage of its total mass will it lose in the next 4.5 billion years (the approximate remainder of its lifespan)?

25.4.9 Determine everything you can Two stars form a binary star system. Each has a mass of 10^{30} kg and travels in a circular path of radius 10^7 m as shown. Each star emits primarily yellow light (550 nm wavelength). Astronomers on Earth (located far to the left of the figure) analyze the light coming from this system over time. Determine the value of as many relevant physical quantities as you can.

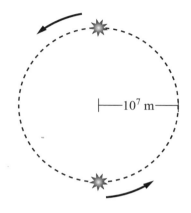

26 Quantum Optics

26.1 | Qualitative Concept Building and Testing

26.1.1 Observe and explain Rub the top of an electroscope (see the sketch at right; rubbing is necessary to "transfer" more charge from the charged object to the electroscope) with differently charged objects. Note what happens to the electroscope leaves in each case. Provide an explanation for the outcome of each experiment.

The experiment	Outcome of experiment	Explanation
a. Rub with a negatively-charged foam pipe. Then remove the pipe. (The foam pipe is charged negatively if rubbed with wool.)	The leaves remain deflected for a long time.	
b. Rub with a positively-charged foam pipe. Then remove the pipe. (The foam pipe is charged positively if rubbed with plastic wrap.)	The leaves remain deflected for a long time.	
c. Rub with a negatively-charged foam pipe. Remove the pipe. Then shine a flashlight on the metal disc on the top.	The leaves remain deflected for a long time.	
d. Rub with a negatively-charged foam pipe. Remove the pipe. Then shine an ultraviolet (UV) light on the metal disc.	The electroscope discharges immediately (as indicated by the leaves moving together).	
e. Repeat (c) and (d), only this time use a positively charged foam pipe.	The leaves remain deflected for a long time.	

The discharge of the negatively charged electroscope due to exposure to light is called a *photo-electric effect*.

26.1.2 Explain Assume you know that free electrons inside metals can move and positively charged ions cannot. Assume that light is an electromagnetic wave in which $\vec{E}$ and $\vec{B}$ fields oscillate periodically.

a. Use your knowledge of the effects of the electric field on charged particles to explain the effect of the light on the negatively charged metal surface in Activity 26.1.1d (think microscopically).

b. Repeat for Activity 26.1.1e.

26.1.3 Observe Physicists use an evacuated glass container such as the one shown to study the photoelectric effect. Light of different frequencies can shine through a quartz window onto a metal plate connected to the negative pole of the battery. Such a plate is called the cathode container. When no UV light shines on the cathode, the ammeter does not register any current.

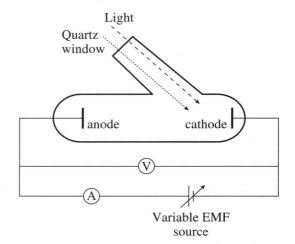

a. When a UV light shines on the cathode, the ammeter registers a current in the circuit. Explain how UV light can cause the current. Note: A voltmeter has very high electric resistance.

b. When the battery is turned off but the UV light still shines on the cathode, the ammeter registers a small current—much smaller than in case (a). Explain.

c. If the polarity of the battery is reversed, then the plate on which the light shines is at a higher potential than the plate on the left side. When this reversed potential difference reaches a certain value, the ammeter stops registering any current. Explain why. (This potential difference is called a stopping potential difference, ΔV_s.)

26.1.4 Observe and explain Referring to the apparatus in Activity 26.1.3, the current is stopped when the metal cathode on which the light shines is at a positive potential relative to the more negative potential of the plate on the left side.

a. Surprisingly, you find that the potential difference that stops the current (see Activity 26.1.3c) does *not* depend on the intensity of light. The electric current induced by high-intensity light is stopped as easily as electric current induced by low-intensity light. Explain this observation.

b. While the stopping potential does not depend on the intensity of light, it does depend on the color: the higher light frequency (UV versus visible, violet versus red), the higher the stopping potential. How can you explain this observation?

26.1.5 Explain You observed in Activity 26.1.1 that visible light does not discharge a negatively charged electroscope. The increase of the intensity of visible light does not make a difference—no current is observed. However, even at very low intensity UV light produces electric current. Explain.

26.1.6 Represent and reason Analyze and represent the following two historical findings:

a. In 1902 the German physicist Phillip Lenard suggested an explanation for the photoelectric effect. He proposed that light being an electromagnetic wave knocked out electrons from the surface of the cathode by continuously exerting force on the electrons. These electrons were then accelerated by the electric field of the battery inside the glass tube, reached the opposite electrode and closed the circuit. He reasoned that if the energy of interaction between electrons and the lattice is negative and equal to $-\phi$, and light had the energy E_{light} larger than ϕ, then the leftover energy of light would be given to the electrons in the form of kinetic energy K_f. Draw a new energy bar chart that represents this energy exchange process between light and electron-lattice system during the photoelectric effect.

b. In 1905 A. Einstein suggested that the photoelectric effect can be explained assuming that light is a stream of bundles of energy (photons), which are individually absorbed by electrons in the metals. The energy of each photon is determined by the frequency of light ($E = hf$, where $h = 6.63 \times 10^{-34}$ J · s); the higher the frequency, the higher the photon energy. An electron is bound to the crystal lattice, and the energy of the interaction of one electron with the lattice is $-\phi$. An electron can absorb only the energy of one photon. Draw a new energy bar chart that represents the energy exchange process between a photon and an electron-lattice during the photoelectric effect.

26.1.7 Explain Use the photon model of light to explain why light exerts pressure on a surface on which it shines. On what surface would the same photon exerts a greater pressure: a shiny one or a black one? Explain. *Hint:* Think about elastic collisions and inelastic collisions.

26.2 | Conceptual Reasoning

26.2.1 Represent and reason Draw an energy bar chart to represent the following process: a photon of light hits a metal and gets absorbed. The energy of the photon is exactly equal to the magnitude of the negative electric potential energy of the interaction between the electron and the lattice.

26.2.2 Represent and reason Draw an energy bar chart to represent the following process: a photon of light hits a metal and ejects an electron with zero kinetic energy.

$$\underline{U_{qi}} \quad + \quad \underline{E_{light}} \quad = \quad \underline{K_{ef}}$$

26.2.3 Represent and reason Draw an energy bar chart to represent the following process: a photon of light hits a metal and ejects a fast-moving electron.

$$\underline{U_{qi}} \quad + \quad \underline{E_{light}} \quad = \quad \underline{K_{ef}}$$

26.3 | Quantitative Concept Building and Testing

26.3.1 Observe and explain: In the table below try to use the *wave model* of light to explain each of the experimental results involving the apparatus shown in Activity 26.1.3.

The experiment	Result	Explain using wave theory.
a. As the light intensity increases, the electric current changes as shown in the graph.	Current vs. Intensity (0)	
b. The dependence of the electric current on the potential difference across the electrodes is shown. Explain the steady part of the graph. The intensity of light remains constant during the experiment.	Current vs. ΔV, V_s, 0. Note: ΔV is positive when the left metal plate (the cathode) connects to the positive battery terminal.	

The experiment	Result	Explain using wave theory.
c. Use the wave model to try to explain why the current decreases to zero when there is a negative stopping potential difference ΔV_s in (b).		
d. You repeat the previous experiment for increasing intensity light. The stopping potential difference ΔV_s does not change.		
e. The potential difference ΔV_s needed to stop the electric current depends on the light frequency— see the graph.		

26.3.2 Observe and explain: Use a photon model to try and explain the results of the five experiments in Activity 26.3.1. Before starting, carefully describe the new model.

The experiment	Result	Explain using the photon model.
a. As the light intensity increases, the electric current changes as shown in the graph.		
b. The dependence of the electric current on the potential difference across the electrodes is shown (measured by the voltmeter) when the intensity of light remains constant. Explain the steady part of the graph.	Note: ΔV is positive when the left metal plate (the cathode) connects to the positive battery terminal.	

(continued)

The experiment	Result	Explain using the photon model.
c. Use the photon model to explain why the current decreases to zero when there is a negative stopping potential difference ΔV_s in (b).		
d. You repeat the previous experiment for increasing intensity light. The stopping potential difference ΔV_s does not change.		
e. The potential difference ΔV_s needed to stop the electric current depends on the light frequency—see the graph.	ΔV 0 ——— f_{cutoff} ——— f	

26.3.3 Test your ideas:
Light passes through double slits and illuminates a screen producing the double-slit interference pattern observed and explained in Chapter 23 using a wave model of light. However, we now have experiments that can only be explained if we consider light to be a stream of photons—a photon model for light.

a. First use the wave model of light to predict what you expect to observe on the screen if you illuminate the double slits with very low intensity light. Then use the photon model of light to make a prediction.

b. Reconcile these two models of light with the outcome of this experiment, which is described in the textbook in Section 26.4.

26.3.4 Evaluate
Your friend Mark says that the photon model of light is not a new model, but just an old particle model of light. How can you convince Mark that his opinion is not correct?

26.3.5 Derive We found that light has a particle-like behavior when interacting with matter. If a photon is particle-like, it should have momentum.

a. Write an expression for the energy of a photon and set it equal to the relativistic energy of a particle with mass m moving at speed c. (Massive particles do not move at the speed of light, c. Here we are assuming that the photon is a particle moving at speed c.) Use this to determine an expression for the equivalent mass of a photon.

b. Write an expression for the momentum of a photon—it moves at speed c and has the equivalent mass m derived in part (a). You should now have an expression for the momentum of a photon in terms of its frequency.

c. Rewrite this expression in terms of the wavelength of the photon.

d. Compare a photon to a classical particle—such as a billiard ball. What are the properties that are similar? What are the properties that are different?

26.3.6 Represent and reason: Find v_2 when a moving billiard ball collides elastically with a stationary ball.

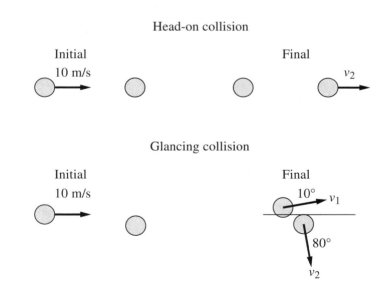

26.3.7 Test your idea Use concept of the momentum of a photon to analyze the following process. A photon of wavelength λ scatters off a particle of mass m that is at rest. The photon moves off at angle θ and the particle at angle α. In addition to the change in the direction of travel, what else should happen to the photon? Use momentum bar charts to analyze the collision; remember that momentum is a vector quantity and you need to make two component bar charts.

26.4 | Quantitative Reasoning

26.4.1 Observe and explain: The experiment described in Activities 26.3.1(e) and 26.3.2(e) is repeated for different types of metal. The results are shown at the right. The work functions of several metals are given in the table below. Use the photon model of light to explain these observations.

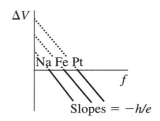

Metal	Aluminum (Al)	Copper (Cu)	Iron (Fe)	Sodium (Na)	Platinum (Pt)	Zinc (Zn)
Work function (eV)	4.1	4.7	4.7	2.3	6.4	4.3

26.4.2 Practice Use some of the information from Activity 26.4.1 to determine the minimum frequency light that will cause a photoelectric emission (a) from sodium and (b) from iron.

26.4.3 Represent and reason A 200-nm light source shines on a sodium surface. Represent with a bar chart and an equation each of the processes described below.

Word description of a process	Energy bar chart description of the process	Mathematical description of the process
a. The process starts with a 200-nm photon and ends just after the ejection from the sodium of an electron that moves at maximum speed.		
b. The same as above, only now the electron has traveled across the photoelectric tube. A potential difference has stopped the electron just before reaching the metal collector electrode.		

26.4.4 Represent and reason: Draw an energy bar chart for each of the three photoelectric effect processes described below. Then write a mathematical description for each process (if it can occur).

Word description of a process	Energy bar chart description of the process	Mathematical description of the process
a. The process starts with a photon whose energy is $hf > \phi$ and ends just after the ejection of an electron from a metal surface while the electron moves at maximum speed.		
b. The same as above only now the energy of the photon is $hf = \phi$.		
c. The same as above only now the energy of the photon is $hf < \phi$.		

26.4.5 Evaluate Your friend Tamara is working on physics problems. She provides the following answer to the problem below. Evaluate the answer to see if you agree. If not, correct the answer.

The problem:

a. What is maximum speed of electrons leaving the copper cathode of work function 4.7 eV? The stopping potential difference is 3.0 V.

b. What is the wavelength of this light?

Proposed solution:

a. The electron's kinetic energy must have been enough to traverse a region with a $(4.7 + 3.0)$ V potential difference. Thus, using energy conservation we get:

$$\frac{1}{2}mv^2 = e\Delta V$$

or

$$v = \left(\frac{2e\Delta V}{m}\right)^{1/2} = \left(\frac{2(1.6 \times 10^{-19}\,\text{C})(7.7\,\text{V})}{(9.11 \times 10^{-31}\,\text{kg})}\right)^{1/2} = 1.64 \times 10^6 \text{ m/s.}$$

The photon's energy equals the kinetic energy that the electron acquired from it, which equals the stopping energy of the electric potential difference:

$$hf = \frac{1}{2}mv^2 = (-e)(-V_s)$$

b. Note that $f = c/\lambda$. Thus,

$$\lambda = \frac{hc}{eV_s} = \frac{(6.63 \times 10^{-34}\,\text{J}\cdot\text{s})(3.0 \times 10^8\,\text{m/s})}{(1.6 \times 10^{-19}\,\text{C})(3.0\,\text{J/C})} = 414 \times 10^{-9}\,\text{m} = 414\,\text{nm}.$$

26.4.6 Explain You have a laser pointer. Remember that lasers emit monochromatic light—light having a single frequency.

a. How is the color of the laser beam related to the energy of the photons?

b. How is the intensity of the light (energy/time) related to the number of photons per second?

c. How is the intensity of the light (energy/time) related to the frequency of the photons?

27 Atomic Physics

27.1 | Qualitative Concept Building and Testing

27.1.1 Observe and explain In the late 19th century, several observational facts had been established that any model of the atom had to explain. One included observations of light emitted by low-density gases. The experimental set up for these observations looks as follows: A glass tube filled with gas has two metal electrodes in it. The electrodes are connected to the poles of a high-voltage power supply. When the power is on, the gas in the tube glows. Different gases glow with different colors.

a. Use a spectroscope (or a simple grating) to examine light emitted by a tube filled with hydrogen. Describe your observations.

b. Compare your observations to observations of the light of a bright lightbulb filament when viewed through the spectroscope.

27.1.2 Observe and explain To help determine atomic structure, Philip Lenard shot cathode rays (electrons) at a thin sheet of aluminum. He observed that electrons moved through the foil without any deflection. At this time physicists already knew that atoms are electrically neutral and contain several electrons. Electrons were known to have a very small mass, much smaller than the mass of the atom. An atomic model developed by J. J. Thomson included positive charge equally distributed within the atom with electrons embedded in it—like plums in a positively-charged plum pudding. Explain how this model accounted for Lenard's experiments.

27.1.3 Observe and explain Alpha particles are elementary particles emitted by some materials, such as uranium; an alpha particle has an electric charge of $+2e$ and the mass of a helium atom.

To help determine atomic structure, Rutherford and his graduate students shot alpha particles at a thin sheet of mica or metals. They looked at places where the particles hit a fluorescent screen after passing through the thin sheets. They also looked for alpha particles that possibly bounced backward off the thin sheet (on the same side as the incoming alpha particles). They observed that most alpha particles moved through the sheet with minor deflection. However, a small fraction of the alpha particles bounced backward. At this time physicists already knew that atoms are electrically neutral and contain several electrons. Electrons were known to have a very small mass, much smaller than the mass of the atom. What could Rutherford suggest about the distribution of the positive charge in the atom to explain his experiments?

27.1.4 Design an experiment Three solid discs are secured below and hidden under a board above. The discs of diameter d are spread randomly across a distance L, as shown at the right. You obtain BBs (ammunition for BB guns) that have a diameter much smaller than diameter d. You roll these small BBs through the opening under the boards and observe that some of the BBs are scattered to the sides and even backwards. Many BBs move straight ahead with no scattering. Design an experiment and carefully describe how you can estimate the diameter of the hidden discs by this BB scattering experiment. You can roll the BBs many times.

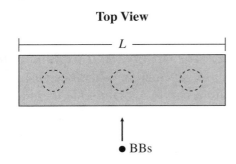

Top View

27.1.5 Explain Using similar ideas Rutherford estimated the size of a positively charged atomic nucleus to be about 10^{-15} m. At that time people already knew the approximate size of atoms to be about 10^{-10} m. How could he reconcile these two numbers? What possible model of the atom could he suggest that would explain Thomson's cathode ray experiments, Lenard's experiments, the experiments of his colleagues, and the fact that atoms are neutral and stable?

27.1.6 Explain Does the planetary atomic model proposed by Rutherford and information about the radiation of the electromagnetic waves that you learned in Chapter 24 (any accelerated electrically charged particle emits electromagnetic radiation):

a. Produce a stable model for the hydrogen atom with one electron circling around the nucleus? Explain.

b. Explain the lines in a spectrum of hydrogen (see Activity 27.1.1).

27.1.7 Evaluate In 1913 Niels Bohr devised a solution to save Rutherford's planetary model from collapsing and to explain the line spectrum of hydrogen. He suggested that electrons are charged particles, which when bound in an atom behave in this way: they obey Newton's laws, interact with the nucleus via Coulomb forces, and do not radiate energy when the atoms are in the preferred energy states (called stable energy states) even though electrons in those atoms are moving in a circle. But as a trade off, atoms can only have energies corresponding to the energies of the stable states. To change its state, the atom needs either to emit a photon (when the energy decreases), or to absorb a photon (when the energy increases). Discuss how this model helps to explain observations of gas spectra and observations conducted by Rutherford and his colleagues.

27.1.8 Observe and explain Imagine a cathode ray tube with a very hot cathode. Electrons in the metal wire have considerable kinetic energy and escape the cathode and accelerate toward the positively charged anode. Some of these electrons traveling at about the same speed pass through an opening in the anode and then reach two closely-spaced narrow slits. A fluorescent screen beyond the slits indicates places where electrons hit the screen. The pattern of electrons that hit the screen is shown below—first only a few electrons, and later after many have hit it. Develop two or more explanations for this observed pattern.

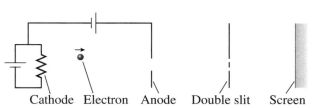

Cathode Electron Anode Double slit Screen

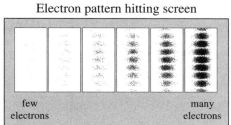

Electron pattern hitting screen

few electrons   many electrons

27.1.9 Observe and explain You repeat the experiment described in Activity 27.1.8 only now you increase the electric potential difference from the cathode to the anode so that the electrons are moving faster when they reach the two slits. You find that the pattern on the screen changes in a systematic way—the faster the moving electrons, the closer the spacing of the bands of electrons hitting the screen. Try to account for this observation.

27.2 | Conceptual Reasoning

27.2.1 Reason Discuss different aspects of the energy of an electron-nucleus system in a hydrogen atom in Bohr's model. Fill in the table that follows.

Does the system posses kinetic energy? Explain.	Does the system posses electric potential energy? Explain.	Does the system possess gravitational potential energy? Explain.	Draw an energy bar chart representing one energy state of the system. Decide on the scale and the direction of the bars.

27.2.2 Explain Explain the observations of light emitted by gases in tubes using the photon model of light. What do the colors of the lines correspond to? What does the brightness of the lines correspond to?

27.2.3 Reason In Bohr's model of the atom, the electron revolves around the nucleus similar to the planets revolving around the Sun (the solar system is an analogy for the atomic system). To understand the model better, answer the questions about the analogy in the table below.

Find the analogous objects in two systems.		Find analogous interactions in two systems.		Find the aspects of the atomic model that do not have analogies in the solar system model.
Atomic model	Solar system	Atomic model	Solar system	Atomic model only

27.2.4 Test your idea An electron beam such as described in Activity 26.1.6 passes through a grating instead of double slits.

a. You keep the speed of the electrons fixed but use different gratings that have first few slits per centimeter and then other gratings with increasing numbers of slits per centimeter. Predict how the pattern of electrons hitting the screen beyond the gratings varies. Give reasons for your predictions.

b. Next you keep the grating fixed (using the one with the most slits per centimeter) and now vary the electric potential from the cathode to the anode. Qualitatively, predict how you think the pattern of the electrons hitting the screen changes and explain carefully the reasons for your predictions.

c. Compare and contrast a photon beam with an electron beam. What are the similar properties of both beams? What are the properties that only one beam possesses?

27.3 | Quantitative Concept Building and Testing

27.3.1 Reason Bohr's postulates suggest that the frequency of the emitted photons can be found using the following mathematical expression: $U_{a2} - U_{a1} = hf$, where U_{a2} is the energy of the atom in state 2, U_{a1} is the energy of the atom in state 1, and hf is the energy of the emitted photon.

a. Does this relation make sense in terms of units? Explain.

b. Does it make sense in terms of the observations of light emitted by low-density gases? Explain.

c. Does this expression make sense in terms of light emitted by the lightbulb filament? Explain.

27.3.2 Reason Bohr's postulates suggest that an the atom can only be in the states where the product of the electron's mass, speed, and distance from the nucleus, mvr (its orbital angular momentum) equals a positive integer times $h/2\pi$.

a. Does this expression make sense in terms of units? Explain.

b. What does this expression imply about the velocity of the electron in different energy states?

27.3.3 Derive Use Bohr's postulates and your knowledge of electrostatic interactions and circular motion to find the value of the smallest electron orbit in the hydrogen atom.

a. What is the electric potential energy of the electron–nucleus system? Is it positive or negative? Assume that the atom is neutral.

b. What is the kinetic energy of the system?

c. What is the total energy of the system?

d. How is the velocity of the electron related to the distance between the electron and the nucleus? Use your knowledge of circular motion and Bohr's third postulate to answer this question.

e. Combine the results of (a) through (d) to determine the smallest-radius electron orbit in a hydrogen atom (the atom is said to be in the ground state).

f. Evaluate your result.

27.3.4 Derive Using Bohr's postulates and the relationships that are assumed to be valid to describe the hydrogen atom (namely Coulomb interactions between the electron and the nucleus, the circular motion of the electron, and Newton's second law), develop a plan to determine the wavelengths of bright lines that you can see when looking at the hydrogen tube through a spectrometer. Do not calculate anything yet. Remember that a human eye is sensitive to the wavelengths of 300–800 nm.

27.3.5 Test your idea Use the plan that you outlined in Activity 27.3.4 to predict the wavelengths of visible light emitted by hydrogen gas.

a. What energy transitions are occurring to produce these wavelengths?

b. Compare the spectrometer lines to your predictions. Were the predictions matched?

27.3.6 Reason Imagine that a hydrogen atom is in its ground state.

a. What is the total energy of the atomic system?

b. In part (a) you obtained a negative value. Does it make sense? Explain.

c. What is the minimum energy of the photon that a ground-state hydrogen atom needs to absorb for the electron to become free? Explain.

d. Express the value that you obtained in part (c) in the units of electron volts. One electron volt is the energy that an electron acquires when it passes through a potential difference of 1 V. The magnitude of the charge of the electron is 1.6×10^{-19} C.

27.3.7 Observe and explain
Using a variety of detectors one can observe spectral lines from hydrogen atoms in the visible, ultraviolet and infrared parts of the electromagnetic spectrum. The series of four lines in the visible part were called the Balmer series with wavelengths of 656.21 nm; 486.07 nm; 434.01 nm, and 410.12 nm. A series of lines in the infrared part of the spectrum (3 lines) were called the Paschen series. And finally, the most recently discovered series in the UV part of electromagnetic wavelengths (4 lines) were called the Lyman series. The names of the series are the names of the physicists who found the lines. They also found that the wavelengths of light for the lines in all series can be empirically described by the formula

$$1/\lambda = R_H\left(\frac{1}{n_f} - \frac{1}{n_i}\right).$$

a. To understand the meaning of the symbols in the empirical formula, fill in the table.

Determine n_i and n_f for the Balmer series.	Determine the value of the constant R_H using the knowledge of the wavelengths of the Balmer series.	Determine n_i and n_f for the Lyman series.	Determine the wavelengths of the 4 lines in the Lyman series.	Determine n_i and n_f for the Paschen series.

b. Derive the empirical formula using the knowledge of Bohr's model of the atom.

27.3.8 Represent and reason
In the figure at the right the hydrogen atom is represented as a series of spheres corresponding to the possible orbits of the electron corresponding to the allowed energies. The orbit with $n = 1$ is called a ground state. Use this representation to answer the following questions:

a. What is the approximate scale of the picture, i.e., how large is the radius of the $n = 1$ sphere?

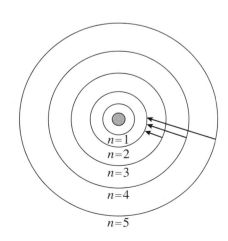

b. Where in this picture is the electron when the atom has the smallest energy? The largest energy?

c. Which series of spectral lines do the arrows represent? How do you know?

d. Draw similar arrows corresponding to the other two series.

27.3.9 Observe and explain Imagine using a set up as in the Activity 27.1.6. Only this time light emitted by a hot lightbulb filament passes through a container with cold atomic hydrogen. A person observes the light though a spectrometer. She sees a colored band in which red color slowly turns into orange, then into yellow, then into green, into blue, and finally into violet (continuous spectrum) with dark lines. The wavelengths at which the dark lines appear are equal to: 656.21 nm; 486.07 nm; 434.01 nm, and 410.12 nm. Explain why a person sees dark lines at these particular wavelengths. Assume that the lightbulb radiates light of all wavelengths.

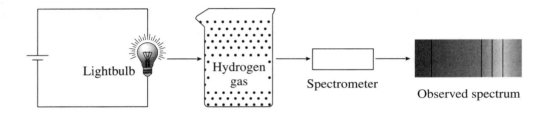

27.3.10 Reason Electrons could be described as having wave-like properties. Their wavelength can be determined using the relationship $p = \dfrac{h}{\lambda}$. Then one can suggest that if an electron has a wavelength that allows it to form a standing wave while orbiting the nucleus, then this state will be a stable state.

a. Show how this idea can explain Bohr's third postulate. *Hint:* Think of what wavelength an electron's standing wave can have when it occupies a certain circular orbit.

b. Discuss how this idea changes Bohr's model of the atom.

27.3.11 Derive A small atomic-size particle (like an electron or proton) of mass m is moving with speed v.

a. Write an expression for the momentum of this particle. Set this expression equal to the expression for the momentum of a photon (the expression in terms of the wavelength associated with the photon).

b. Solve this equation for an expression for the wavelength of the moving atomic-size particle.

27.4 | Quantitative reasoning

27.4.1 Reason Determine the temperature of the hydrogen gas at which most of its atoms are ionized. Indicate all assumptions that you made.

27.4.2 Represent and reason Use the knowledge of allowed energy states of a hydrogen atom to draw a scaled energy diagram for the hydrogen atom.

27.4.3 Represent and reason Complete the energy diagrams to represent each of the following processes:

An atom absorbs a 15 eV photon.	An atom in the state with $n = 5$ changes into the state with $n = 2$.	An atom absorbs a photon and changes from the state with $n = 1$ to the state with $n = 4$ and then after 10^{-8} s changes to the state with $n = 2$.	Hydrogen gas is placed in the electric field so the potential difference between the ends of the container slowly changes.	Hydrogen gas surrounds a hot object whose temperature is above 10^6 K.

27.4.4 Reason Consider the following process: a hydrogen atom absorbs a photon and changes its energy from the second state to the fifth state. Complete the table that follows.

What is the energy of the atom in the second state?	What is the energy of the atom in the fifth state?	What is the energy difference between the states?	What is the energy of a photon that the atom must absorb to undergo this change?	What is the frequency of the photon that the atom must absorb to undergo this change?

27.4.5 Reason Imagine that you have an atom of He that lost one electron and is now a positively charged ion with the charge of the nucleus equal to $+2$ (He^+).

a. Determine the radii and energy of the $n = 1, 2, 3$ energy states in this ion.

b. Construct an energy-level diagram for this ion.

27.4.6 Reason In Bohr's model of the hydrogen atom the electron revolves around the nucleus like a planet revolves around the Sun. Determine the speed and frequency of the revolution of an electron around the first Bohr orbit in hydrogen. According to classical physics, the atom should emit electromagnetic radiation at this frequency (because circular motion described mathematically is very similar to vibrational motion). In what portion of the electromagnetic spectrum is this frequency?

27.4.7 Reason Niels Bohr proposed the postulates that described the behavior of the electron in the hydrogen atom as different from electrons that are not bound to nuclei. However, he also stated that "At sufficiently large n, the new model should predict the same behavior of the electron as classical physics." Discuss what this statement means and whether Bohr's model of the atom behaves according to this principle.

27.4.8 Regular problem A hydrogen atom changes its state from $n = 15$ to $n = 5$. Complete the table that follows.

Represent the process with an energy diagram.	Discuss whether the energy of the system increases or decreases.	Discuss whether the atom absorbs or emits a photon to undergo this process.	Calculate the energy, the frequency, and the wavelength of the emitted photon.

27.4.9 Reason The average thermal energy of the random translational motion of a hydrogen atom at room temperature is $(3/2)kT$ where k is the Boltzmann constant. Would a typical collision between two hydrogen atoms be likely to transfer enough energy to one of the atoms to cause a transition from the $n = 1$ energy state to the $n = 2$ state?

a. Draw an energy bar chart for a process during which two hydrogen atoms collide and one of them changes state from $n = 1$ to $n = 2$.

b. Would a typical collision between two hydrogen atoms in Earth's atmosphere be likely to transfer enough energy to one of the atoms to change the energy state from $n = 1$ to $n = 2$? Explain your answer. (*Note*: Actually, free hydrogen on Earth is in the molecular form H_2. However, the above reasoning still explains why a gas composed of hydrogen molecules at room temperature is seldom excited to the point of emitting light.)

c. Find the temperature at which the collisions between hydrogen atoms can lead to this change of state.

27.4.10 Regular problem A gas composed of hydrogen atoms in a tube is excited by collisions with free electrons. If the maximum excitation energy gained by an atom is 12.5 eV, determine the wavelengths of light emitted from the tube as atoms return to the ground state.

27.4.11 Regular problem Determine the de Broglie wavelength of an electron that has been accelerated across a potential difference of 200 V. The electron is initially at rest and has a mass of 9.1×10^{-31} kg.

27.4.12 Regular problem

a. Determine the states of the four electrons in the ground state of beryllium.

b. Determine the electron configuration of the twelve electrons in the ground state of magnesium.

27.4.13 Design an experiment
Design and describe an experiment to determine whether gas collected in a cave contains carbon monoxide (CO).

28 Nuclear Physics

28.1 | Qualitative Concept Building and Testing

28.1.1 Observe and explain Photographic plates are glass plates covered with material that undergoes a chemical reaction when light shines on it – this is called "exposure". After the photographic plate undergoes another chemical process called "development" the places on the plate exposed to light change color. In 1896 Henri Becquerel experimented with uranyl crystals and found that when placed on top of a photographic plate the crystals could expose the plate, even when the materials were stored in a dark box. Suggest an explanation for this observation.

Hint: Some phosphorescent or fluorescent materials, such as barium sulfide, zinc sulfide, etc., can expose photographic paper after they have been exposed to light. However, uranyl salts do this without exposure to light.

28.1.2 Test your idea One if the explanations of Becquerel's experiments is that uranyl salts emit some kind of charged particles that affect the photosensitive paper in a way similar to light. How can you test this explanation?

28.1.3 Observe and explain Pierre Curie invented an electrometer that could be used to measure how much charge the electrometer loses per unit time (the current). Marie Curie used this electrometer to measure the amount of electric current produced when the electrometer was placed near uranium salts.

- She found that the amount of current was proportional to the amount of uranium present.
- She found that the intensity of the current was independent of the identity of the uranium salt, its wetness, temperature, physical appearance, or the amount of light shining on it.

a. Explain why the electrometer would lose its charge when a sample of uranium salt was placed nearby.

b. If you were Marie Curie, with a strong background in chemistry, what could you conclude about how the uranium rays are produced? (*Hint:* Sometimes in science you can determine what a phenomenon *is not* long before you have an idea about what it *is*.)

28.1.4 Observe and explain In 1899 Ernest Rutherford and his colleagues investigated the ability of uranium salts to ionize air. He set up two parallel plates, with a potential difference between them. When a uranium sample was placed between the plates, ions created by the radiation would be pulled to the plates before they could recombine. This caused a detectable current. Covering the uranium sample with thin aluminum sheets decreased the amount of current observed, but only up to a point. After this point, no further decrease in current was observed, even with the addition of more aluminum plates. Propose an explanation of why the current decreased with more aluminum shielding, but only to a point.

28.1.5 Explain In 1903 Rutherford placed his radioactive sample in a magnetic field in an apparatus such as shown below. He and his assistants used a scintillating screen, which glowed when a charged particle hit the surface (similar to the screen of an old-fashioned TV that had a cathode-ray tube inside). In the second experiment they used photographic paper and found that it was exposed around point O. Describe below the cause of each exposure.

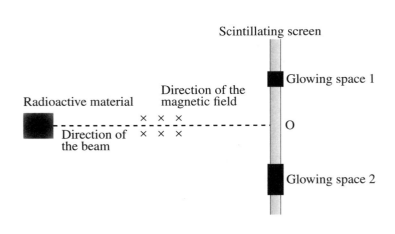

a. Describe everything you can about what caused the glowing screen at space 1.

b. Describe everything you can about what caused the glowing screen at space 2.

c. Describe everything you can about what caused the photographic paper to be exposed at O.

28.1.6 Evaluate reasoning Based on the experiments such as described above, scientists proposed that the nucleus of an atom is made of positively charged alpha particles and negatively charged electrons. Their electrostatic attraction holds them together. When a nucleus has a lot of alpha particles, they start repelling each other and are likely to leave the nucleus (alpha decay). This leaves too many electrons inside that repel each other and thus electrons are emitted (beta decay). After each transformation the nucleus is left in an excited state and emits a high-energy photon—a gamma ray (gamma decay). Describe how this proposal is consistent with the experiments in Activity 28.1.5.

28.1.7 Observe and find a pattern Frederic Soddy (1913) collected the following data related to the radioactive transformation of uranium. The first product that appears in the sample is thorium, then protoactinium, then another isotope of uranium, and so on. Examine the series of the transformation found by Soddy and explain it using your knowledge of alpha and beta decays. Discuss what quantities are conserved in each process. In the series presented below the left subscript indicates the positive charge of the nucleus in the units of the electron charge and left superscript indicates the mass of the nucleus in the atomic units. Using this notation system hydrogen can be written as $^{1}_{1}\text{H}$.

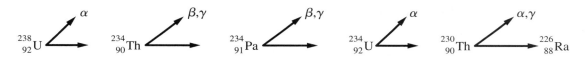

28.1.8 Explain According to Heisenberg's uncertainty principle, the uncertainties of the position Δx and momentum Δp_x of an atomic-size particle can be known no better than $\Delta x \cdot \Delta p_x \geq h/4\pi$, where h, called Planck's constant, equals 6.63×10^{-34} J · s. Experiments by Rutherford's colleagues led to the estimation of the size of atomic nucleus to be about 10^{-15} m. Explain how the application of the uncertainty principle indicates why electrons could not be inside a nucleus.

28.1.9 Explain In 1920 Rutherford suggested that experimental evidence about nuclei at that time indicated the need for a particle that has no electric charge and has a mass slightly higher than the proton nucleus of hydrogen. Irène Joliot-Curie (daughter of Marie Curie and Pierre Curie) and her husband, Frédéric Joliot-Curie, performed experiments in which they found that fast moving alpha particles passing through and being stopped by targets made of light elements (for example, beryllium) created radiation that could penetrate materials better than gamma rays. James Chadwick explained the results of their experiments by using the idea of the particle predicted by Rutherford—a neutron. Use this information to answer the following questions.

a. Use the concepts of protons and neutrons to explain the atomic number and the mass of nuclei.

b. Use the concepts of protons and neutrons to explain alpha decay.

c. Use the ideas of protons and neutrons to explain the beta decay.

d. Suggest an explanation for the fact that positively charged protons and electrically neutral neutrons stay together in a nucleus.

28.1.10 Terminology Determine the number of protons and neutrons in each of the following nuclei:

$^{11}_{5}B$	$^{19}_{9}F$	$^{39}_{19}K$	$^{63}_{29}Cu$	$^{138}_{56}Ba$	$^{208}_{82}Pb$
protons:	protons:	protons:	protons:	protons:	protons:
neutrons:	neutrons:	neutrons:	neutrons:	neutrons:	neutrons:

28.1.11 Observe and find a pattern Devise one or more rules that seem necessary in order for the nuclear reactions below to occur.

a. $^{1}_{0}n + {}^{235}_{92}U \rightarrow {}^{147}_{56}Ba + {}^{86}_{36}Kr + 3\,{}^{1}_{0}n + $ energy

b. $^{239}_{94}Pu \rightarrow {}^{235}_{92}U + {}^{4}_{2}He + $ energy

c. $^{191}_{76}Os \rightarrow {}^{191}_{77}Ir + {}_{-1}^{0}e + $ energy

28.2 | Conceptual Reasoning

28.2.1 Estimate Use what you have learned to estimate the following values.

a. Estimate the density of a nucleus. Indicate any assumptions you made.

b. Estimate Earth's radius if it had this same density. Indicate any assumptions you made.

c. Estimate the radius of a sphere that would hold all of your body mass if its density equaled the density of a nucleus. Indicate any assumptions you made.

28.2.2 Estimate Use what you have learned to estimate the following quantities.

a. Estimate the total number of nucleons (protons and neutrons) in your body.

b. Estimate the total number of electrons in your body.

c. Indicate roughly the volume in cm^3 occupied by these nucleons.

28.2.3 Reason Insert the missing symbol in the following reactions.

a. ${}^{4}_{2}He + {}^{12}_{6}C \rightarrow {}^{15}_{7}N + ?$

b. ${}^{2}_{1}H + {}^{3}_{1}H \rightarrow {}^{4}_{2}He + ?$

c. ${}^{1}_{0}n + {}^{235}_{92}U \rightarrow {}^{140}_{54}Xe + ? + 2\,{}^{1}_{0}n$

d. ${}^{2}_{1}H \rightarrow ? + {}^{0}_{-1}e$

28.2.4 Reason The following nuclei produced in a nuclear reactor each undergo radioactive decay. Write the decay reaction equation.

a. ${}^{239}_{94}Pu$ alpha decay:

b. ${}^{144}_{58}Cs \; \beta^-$ decay:

c. ${}^{65}_{30}Zn \; \beta^+$ decay:

28.3 | Quantitative Concept Building and Testing

28.3.1 Observe and find a pattern Rutherford found that by blowing air across a sample of thorium, a radioactive gas could be collected. This gas lost its radioactivity rapidly, as shown in the table. Is there a mathematical trend in Rutherford's data? Summarize it in your own words.

Time (min)	Portion of time-zero radioactivity lost
0	0
1	~50 %
2	~75 %
3	~87 %
4	~93 %
5	~97 %
10	Undetectable ≈ 100 %

28.3.2 Observe and explain Measurements with a mass spectrometer indicate the following particle masses: $_2^4\text{He}$ (4.002604 u), $_1^1\text{H}$ (1.007825 u), and $_0^1 n$ (1.008665 u). Compare the mass of the helium atom to the mass of the particles of which it is made. What do you conclude? *Note:* 1 u is $1/12$ the mass of a carbon-12 atom and equals 1.660566×10^{-27} kg $= 931.5$ MeV$/c^2$.

28.3.3 Reason in Activity 28.3.2 you found that the mass of a helium nucleus is less than the mass of the nucleons inside it.

a. Explain how this observation led scientists to the idea that it is possible to convert hydrogen into helium to produce thermal energy.

b. Does this process mean that energy is not conserved? Explain.

c. Represent the process with an energy bar chart. What is the binding energy on this chart?

d. Explain why very high temperatures and pressures are needed for this reaction.

e. Estimate the temperature at which two protons will join together due to their nuclear attraction. Remember, that nuclear forces are effective at distances less than or equal to about 10^{-15} meters. Hint: Use an energy approach and not a force approach.

28.3.4 Explain Explain why the following reaction does not occur spontaneously: $^{4}_{2}\text{He} \rightarrow ^{3}_{1}\text{H} + ^{1}_{1}\text{H}$.

28.3.5 Reason In the 1940s Lise Meitner, Otto Hahn, and Fritz Strassmann irradiated uranium with neutrons. They found that instead of getting a heavier uranium isotope, the reaction produced lighter nuclei, like isotopes of Ba (barium). How can you explain their findings?

28.3.6 Reason Lise Meitner asked her nephew Otto Robert Frisch to help with the explanation. They thought about Bohr's liquid drop model; heavy nuclei behaved like a drop of water with a kind of "surface tension" holding it together. The only problem was that the nucleus had an electric charge that would counteract the effect of the surface tension, especially if the nucleus was not very spherical. The model suggested that the nucleus could elongate and divide into two smaller pieces. This meant that the uranium nucleus would be very unstable and ready to divide with the slightest provocation. This could happen when a neutron hit it.

a. Explain how Meitner's and Frisch's reasoning could explain the findings in Activity 28.3.5.

b. Use the liquid drop model to predict another product or products of uranium disintegration. If a chemist finds such products in the mixture resulted from the irradiation of uranium with neutrons, then the water drop model is a productive model to explain the observations in Activity 28.3.5.

c. Explain how the model of Meitner and Frisch can be used to explain why uranium can be used as a source of thermal energy. Additional information: the products of the irradiation of uranium with neutrons included not only chemical elements in the middle of the periodic table but also extra neutrons.

28.4 | ■ Quantitative Reasoning

Note that $1 \text{ u} = 1.6606 \times 10^{-27} \text{ kg} = 931.5 \text{ MeV}/c^2$.

28.4.1 Reason Determine the binding energies per nucleon for $^{238}_{92}\text{U}$ and for $^{120}_{50}\text{Sn}$. Based on these numbers, which nucleus is more stable? Explain.

28.4.2 Reason One part of the carbon-nitrogen cycle that provides energy for the Sun is the reaction $^{12}_{6}\text{C} + ^{1}_{1}\text{H} \rightarrow ^{13}_{7}\text{N} + 1.943 \text{ MeV}$. Using the known masses of ^{12}C and ^{1}H and the results of this reaction, determine the mass of ^{13}N.

28.4.3 Reason Determine the missing nucleus in the following reaction and the energy released in the reaction: $^{232}_{92}\text{U} \rightarrow ? + ^{4}_{2}\text{He} + \text{energy}$.

28.4.4 Reason Radon-222 ($^{222}_{86}\text{Rn}$) is released into the air during uranium mining and undergoes alpha decay to form $^{218}_{84}\text{Po}$ of mass 218.0089 u. Determine the energy released by the decay reaction. Most of this energy is in the form of α particle kinetic energy.

28.4.5 Reason Cesium-137 is a waste product of a nuclear reactor. Determine the fraction of ^{137}Cs remaining in a reactor fuel rod:

a. 120 y after it is removed from the reactor.

b. 240 y after it is removed.

c. 1000 y after it is removed.

Half-Lives and Decay Constants of Some Common Nuclei		
Isotope	**Half-life**	**Decay constant** (s^{-1})
$^{87}_{37}$Rb	4.88×10^{10} yr	4.50×10^{-19}
$^{238}_{92}$U	4.5×10^{9} yr	4.9×10^{-18}
$^{40}_{19}$K	1.28×10^{9} yr	1.72×10^{-17}
$^{239}_{94}$Pu	2.44×10^{4} yr	9.00×10^{-13}
$^{14}_{6}$C	5730 yr	3.84×10^{-12}
$^{226}_{88}$Ra	1602 yr	1.37×10^{-11}
$^{137}_{55}$Cs	30.0 yr	7.32×10^{-10}
$^{90}_{38}$Sr	28.1 yr	7.82×10^{-10}
$^{3}_{1}$H	12.4 yr	1.77×10^{-9}
$^{60}_{27}$Co	5.26 yr	4.18×10^{-9}
$^{131}_{53}$I	8.05 day	9.96×10^{-7}
$^{11}_{6}$C	21 min	5.5×10^{-4}

28.4.6 Regular problem A sample of radioactive technetium-99 of half-life 6 h is to be used in a clinical examination. The sample is delayed 15 h before arriving at the lab for use. Use two methods to determine the fraction that remains.

28.4.7 Regular problem To estimate the number of ants in a nest, 100 ants are removed and fed sugar made from radioactive carbon of long half-life. The ants are returned to the nest. Several days later, it is found that of 200 ants tested, only 5 are radioactive. Roughly, how many ants are in the nest? Explain your estimation technique.

28.4.8 Regular problem A sample from a tree uprooted and buried during the Wisconsin glaciation contains 50 g of carbon when it is discovered.

a. If 1 in 10^{12} carbon atoms in a *fresh* tree sample were carbon-14, how many carbon-14 atoms would be in 50 g of carbon from a fresh tree?

b. Determine the carbon-14 activity of this fresh sample.

c. Determine the age of the buried tree if its 50 g of carbon has an activity of -2.2 s^{-1}.

28.4.9 Regular problem Determine the energy released in the following fission reaction:

$$_0^1 n^- + \ _{92}^{235}\text{U} \rightarrow \ _{56}^{141}\text{Ba} + \ _{36}^{92}\text{Kr} + 3 \ _0^1 n.$$

The masses of the nuclei are: $m_n = 1.0087$ u, $m_U = 235.0439$ u, $m_{\text{Ba}} = 140.9141$ u, and $m_{\text{Kr}} = 91.8981$ u.

28.4.10 Energy from gasoline compared to uranium Use the data provided below to compare these two energy sources.

a. Determine the energy release in MeV of gasoline per molecule of *n*-heptane burned. The molecular mass of *n*-heptane is 100 u, and it releases energy at a rate of 4.8×10^7 J/kg.

b. Determine the ratio of energy released by one uranium-nucleus fission (approximately 200 MeV) and one *n*-heptane molecule combustion.

29 Particle Physics

29.1 | Qualitative Concept Building and Testing

29.1.1 Summarize In your own words, describe the models and reasoning that resulted in the prediction of the existence of antimatter.

29.1.2 Observe and analyze The figure shows the path of a charged particle that was produced when a cosmic ray interacted with the atmosphere. The particle is shown passing through a sheet of lead. Assuming the $\vec{B}$-field points into the page, determine the direction this particle is traveling and the sign of its electric charge.

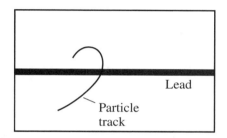

Lead

Particle track

It is determined that the mass of this particle (within experimental uncertainty) is equal to the mass of the electron. Explain why this is interesting.

29.1.3 Observe and find a pattern The table below contains energy and momentum data just before and after a photon undergoes electron-positron pair production.

	Photon	Electron	Positron
Rest energy (MeV)	0	0.511	0.511
Kinetic energy (MeV)	1.62	0.300	0.300
Momentum, x-component (kg · m/s)	8.65×10^{-22}	4.32×10^{-22}	4.33×10^{-22}
Momentum, y-component (kg · m/s)	0	2.50×10^{-22}	-2.50×10^{-22}

Sketch this process.

Find as many patterns as you can in this data. Are they consistent with your understanding of energy and momentum?

29.1.4 Predict and test Describe as many predictions of the Standard Model as you can. For each prediction, was the result of the corresponding experiment consistent or inconsistent with the prediction, or is the result as yet unresolved?

29.1.5 Observe Describe as much evidence as you can in support of the Big Bang model.

29.1.6 Observe Describe as much evidence as you can for the existence of dark matter.

29.1.7 Explain Describe as may hypotheses as you can that explain the nature of dark matter.

29.1.8 Observe Describe as much evidence as you can for the existence of dark energy.

29.1.9 Explain Describe as may hypotheses as you can that explain the nature of dark energy.

29.2 ∎ Conceptual Reasoning

29.2.1 Explain Positron emission tomography can be used in pharmacology to determine if a drug is being delivered to the appropriate places in an organism. Explain how this could work.

29.2.2 Observe and analyze A detector at the Fermi National Accelerator Center (Fermilab) records the collision between a proton and an antiproton. Each gray line in the figure represents the path of a particle created in the collision. Assuming the $\vec{B}$ field points out of the page, determine the sign of the electric charge of each particle. Explain your reasoning.

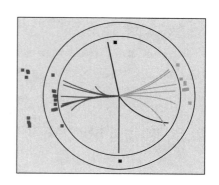

29.2.3 Represent An electron and a positron collide head-on, annihilate, and produce a muon and an antimuon. Sketch a valid version of this process. What assumptions did you make?

Sketch an invalid version of this process. Explain why it is invalid.

29.2.4 Represent Explain why someone might think electron-positron annihilation into a single photon (shown in the figure) is correct.

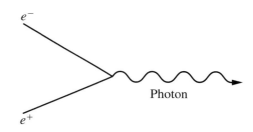

There is a problem with this process, however. Go to the reference frame where the electron and positron are heading toward each other at equal speeds. Draw a sketch of the process from this reference frame.

Is it possible for a single photon to be produced by this annihilation process? Explain your reasoning. Would adding an additional photon help? Explain why or why not.

29.2.5 Represent Explain the following non-fundamental interactions in terms of fundamental ones.

a. The force that air exerts on a moving car.

b. The force that the floor exerts on you while you are standing.

c. The force that one section of a stretched bungee cord exerts on an adjacent section.

d. The force that water exerts on a submerged submarine.

29.2.6 Explain How do we know that the strong interaction has an extremely short range?

29.2.7 Explain Consider the following process:

$$2p \rightarrow 3p + \bar{p}$$

Here the p represents a proton and the $\bar{p}$ represents an antiproton. How is this process possible? Why might someone argue that it is not possible?

29.2.8 Reason Four baryons have been discovered and are known as the delta particles: Δ^-, Δ^0, Δ^+, and Δ^{++}. The superscripts indicate the electric charge of these particles. The delta particles are composed of different combinations of up and down quarks. Determine these combinations.

Given the quark content of the Δ^+, suggest an explanation for what distinguishes it from a proton. *Hint:* Try to relate what you have learned about the hydrogen atom.

29.2.9 Explain Consider the following process:

$$2p \rightarrow p + n + \pi^+$$

Represent this process in terms of the quark content of the particles. Then, explain why this process is an example of confinement.

29.2.10 Explain According to the Big Bang model, how old are the protons and neutrons in your body?

29.2.11 Explain How does the idea of supersymmetry help in explaining dark matter? Dark energy?

29.3 | Quantitative Concept Building and Testing

29.3.1 Analyze How do we know that the electric interaction is much stronger than the gravitational interaction? Why might someone believe the opposite to be true?

Use the interactions between the proton and electron in the hydrogen atom to argue quantitatively why the electric interaction is many orders of magnitude stronger than the gravitational interaction.

29.3.2 Explain You've learned that the electrostatic interaction is weaker the farther apart the two interacting objects are. Here, you will consider how the particle interaction mechanism of fundamental interactions is consistent with this. First, explain this particle interaction mechanism.

Consider two electrons that are very close to each other (say, 10 times the diameter of an atom). Use the uncertainty principle to estimate the energy and momentum of the virtual photons.

Now consider two electrons that are much farther away from each other (say, 1 million times the diameter of an atom). Estimate the energy and momentum of these virtual photons.

Use your above estimates to decide whether the particle interaction mechanism is consistent with your understanding of the electrostatic interaction. Explain your reasoning.

29.4 | Quantitative Reasoning

29.4.1 Reason A photon (gamma ray) of wavelength 1.15×10^{-12} m is produced in a supernova. Is it possible for this photon to produce an electron-positron pair?

Explain your reasoning.

29.4.2 Reason Neutrons are unstable with a half-life of about 15 minutes. They decay into a proton, an electron, and an electron antineutrino. Explain this process in terms of energy, quantitatively.

On the other hand, protons are not observed to spontaneously decay into neutrons, positrons, and electron neutrinos. Explain why.

29.4.3 Regular problem At the Large Hadron Collider protons are collided with each other at extremely high speeds. What is the minimum speed each of two protons must have in order for them to produce one W^+ and one W^- weak interaction mediator? *Hint:* The two protons do not annihilate since they are both protons rather than one proton and one antiproton.